SYSTEMATICS OF THE GENUS HYMENAEA L. (LEGUMINOSAE, CAESALPINIOIDEAE, DETARIEAE)

SYSTEMATICS OF THE GENUS HYMENAEA L. (LEGUMINOSAE, CAESALPINIOIDEAE, DETARIEAE)

BY

YIN-TSE LEE and JEAN H. LANGENHEIM

UNIVERSITY OF CALIFORNIA PRESS
BERKELEY • LOS ANGELES • LONDON

UNIVERSITY OF CALIFORNIA PUBLICATIONS IN BOTANY

Volume 69
Approved for publication November 15, 1974
Issued December 31, 1975

University of California Press
Berkeley and Los Angeles
California

University of California Press, Ltd.
London, England

ISBN. 0-520-09532-4
Library of Congress Catalog Card Number. 74-620192

CONTENTS

SYSTEMATICS OF THE GENUS HYMENAEA L. (LEGUMINOSAE, CAESALPINIOIDEAE, DETARIEAE)

BY

YIN-TSE LEE and JEAN H. LANGENHEIM

dixi itaque hanc Hymenaeam ab Hymenaeo,
deo conjugii veteribus celebrato, cum duo
folia paria conjuncta sint, quae omni nocte,
adhuc dum juniora, connivent, simulque
dormiunt approximata.

Carl von Linné,
Hortus Cliffortianus, 1737

INTRODUCTION

A SYSTEMATIC revision of the genus *Hymenaea* was initiated as part of a coordinated study of the evolution of resin-secreting plants. In a world survey of ambers (Langenheim, 1969) it was pointed out that copious resin production appears to have occurred predominantly in tropical or subtropical environments. *Hymenaea* was chosen as the example for detailed study of a tropical resin-producing genus for several reasons: (1) Tertiary ambers from Mexico, Colombia, and Brazil, derived from *Hymenaea,* provide evolutionary background for the study of chemical changes in the resin; (2) *Hymenaea* is a member of the Leguminosae (Caesalpinioideae), one of the most significant resin-producing groups; (3) the Caesalpinioideae are one of the most important components of equatorial ecosystems in Africa and South America; and (4) *Hymenaea* species are interesting because of their probable origin in African rain-forest ecosystems and adaptive radiation into all major tropical ecosystem types in the New World.

Only a few previous systematic revisions of genera of tropical trees have taken adequate account of ecological and evolutionary considerations. We have placed particular emphasis on our own field collections and ecological observations of the natural populations. Also, we have grown large numbers of seedlings in greenhouses and growth chambers to gain understanding of conditions controlling synthesis and secretion of resins and we have examined specimens from all herbaria where collections were believed to exist. This study is concerned primarily with the analysis of morphological characters, but includes preliminary investigation of reproductive behavior in varying ecosystems which points toward understanding of speciation and of a tentative evolutionary history of the genus.

ACKNOWLEDGEMENTS

The authors are indebted to many individuals in Latin America and Africa for assistance that made possible the field work for this study. Because collections contributing to this work have continued over the period of years from 1967 to the present, on the part of JHL, including those of YTL, it is only possible here to express our appreciation to some of those who have generously aided us.

We are particularly appreciative to the following persons and institutions for their cooperation and hospitality: The late Dr. Faustino Miranda and Dr. Arturo Gomez-Pompa,

Instituto de Biología, Universidad de Mexico, Mexico, D. F.; personnel of the Organization for Tropical Studies, San José, Còsta Rica; Kenyan Forest Service and personnel of East African Herbarium, Nairobi, Kenya; Dr. T. Lasser, Instituto Botánico, Universidad de Venezuela, Caracas, Venezuela; Drs. P. de A. Machado, W. Rodrigues, Instituto Nacional de Pesquisas da Amazonia (INPA), Manaus, Amazonas, Brazil; Dr. M. Pires, Instituto de Pesquisas e Experimentacão Agropecurias do Norte (IPEAN), and Dr. P. Calvancante, Museu Parense "Emilio Goeldi;" Drs. R. Reitz, G. M. Barroso, and D. Sauce, Jardim Botânicō de Rio de Janeiro and Dr. L. E. de Mello Filho, Museu Nacional, Universidade Federal do Rio de Janeiro, Guanabara, Brazil; Drs. C. Bicudo and T. Skvoztzov, Instituto Botânico de Estado, São Paulo, São Paulo, Brazil; Dr. P. de Alvim and Mr. S. G. da Vinha, Centro de Pesquisas do Cacau (CEPEC), Itabuna, Bahia, Brazil; Drs. G. and L. T. Eiten, Departmento de Taxonomia Vegetale, Universidade de Brasilia, Brasilia, D. F., Brazil; and Dr. H. Irwin, New York Botanical Garden, Bronx, N.Y.

We wish to thank Dr. Alva Whittingham, University of California, Santa Cruz, Dr. H. G Baker, University of California, Berkeley, and Dr. Richard Cowan, Smithsonian Institution, Washington, D. C., for their reading of the manuscript and valuable suggestions and comments. We also feel that special recognition and thanks are due to Dr. Dardano Andrade-Lima, Universidade Federal de Pernambuco, Brazil, for his general editorial assistance, and to Dr. Lincoln Constance, University of California, Berkeley, for his unusually careful editing of the Taxonomic Treatment, as well as for their other helpful comments.

Also our gratitude is expressed to personnel of the herbaria who kindly loaned or provided us facilities to examine specimens essential to this study. These herbaria are listed in the text with three exceptions (CEPEC, Centro de Pesquisas do Cacau, SU, Summit Herbarium, Canal Zone, and UP, Herbario Universidad de Panama, Panama) by abbreviations of Lanjouw and Stafleu (1964).

Grateful acknowledgement is made for funds to support this study from The Radcliffe Institute,Organization for Tropical Studies, and National Science Foundation Grants GB-1312, GB-2397, GB-5816, GB-13659, GB-35267, and GB-29278.

TAXONOMIC HISTORY

Hymenaea was one of the genera established by Linnaeus in his *Genera Plantarum* (1737a). The description of the genus, however, was based solely on the accounts and illustrations by Plumier (1703) under the "American" name "Courbaril." The plant, according to Linnaeus (1737b), had been known even earlier as "*Ceratia diphyllos*" by Bauhin (1623), Breyne (1679), Ray (1688), and Plukenet (1696), and under the Brazilian name "Jetaiba" by Piso (1642) and Marcgraf (1648). Linnaeus (1737b) was not pleased with the two native names, indicating they were "barbaric." He chose the name "Hymenaea" referring to Hymenaeus or Hymen, the Greek god of marriage, because the two leaflets are paired and through their nyctitropic movements "lie next to each other at night." *Ceratia,* on the other hand, is considered to be synonymous with *Ceratonia.*

In 1753 Linnaeus described from "Brazil" *H. courbaril,* the type species of the genus. The next species to be described was *H. verrucosa* from Madagascar (Gaertner, 1791). This Madagascan tree had apparently already been mentioned by de Jussieu (1789), who gave it the native name "Tanrojou" without describing the plant. But in 1792 Lamarck described and illustrated a more complete specimen of *H. verrucosa.* Also in 1792 Vahl described *H. venosa* from French Guiana.

In the *Prodromus,* de Candolle (1825a) recorded two more species discovered by Humboldt and Bonpland and described by Kunth (1825) as *H. candolleana* and *H. floribunda.* Then in 1827 Hayne named eight new species of *Hymenaea (H. latifolia, H. confertiflora, H. confertifolia, H. stilbocarpa, H. stigonocarpa, H. olfersiana, H. martiana,* and *H. sellowiana),* but they were not fully described and illustrated by him until 1830. In the same paper Hayne (1827) established the new genus *Trachylobium* and described four species: *T. martianum* (Amazon Basin), *T. hornemannianum* (Mauritius), *T. gaertnerianum* (Java, probably from a botanical garden), and *T. lamarckianum* (unknown locality). Surprisingly, he did not include *H. verrucosa* in the species of *Trachylobium,* although his new genus was based on specimens previously identified as *H. verrucosa.*

Hayne (1830), however, noted:

> Da Gärtner zuerst die *Hymenaea verrucosa* unter welcher mehrere Arten de Gattung *Trachylobium* stecken, beschrieben hat, und ich gerne eine von diesen Arten mit seinem Namen bezeichnen möchte: so wähle ich dazu dies aus Java, da ich die, aus Madagascar herstammende, welche von ihm beschrieben wurde, und vielleicht noch als besondre Art sich unterscheidet, nicht haben konnte. (Hayne, 1830: 1922)

Hayne (1827, 1830) was not concerned with the highly disjunct distribution of *Trachylobium,* resulting from the inclusion of *T. martianum* from Brazil with the other Old World species. However, *T. martianum* was later redefined by Bentham (1870) as *Cynometra spruceana,* resulting in *Trachylobium* being an exclusively Old World genus. In 1862 Klotzsch described, from Mozambique, another species, *T. mossambicense.*

Hayne left two questions unanswered. The first one asked if *Trachylobium* should be considered a genus distinct from *Hymenea.* As early as 1842 Hasskarl returned *Trachylobium* to *Hymenaea* and classified *T. gaertnerianum* as *H. verrucosa.* Following him, Miquel in 1855 cited *T. gaertnerianum* as a synonym of *H. verrucosa* and questioned, "Num species Madagascar *T. lamarckianum* et *T. hornemannianum* Hayne cum hac jungendae?" Dietrich (1840) placed all *Trachylobium* species under *Hymenaea,* whereas Baillon (1870) actually considered *Trachylobium* a section of *Hymenaea.** Cordemoy (1900) and Sornay (1913) also treated *Trachylobium* species under *Hymenaea.*

The second question was concerned with the distinctness of the species included within *Trachylobium.* After reviewing the taxonomic history of the genus (Bentham and Hooker, 1865; Oliver, 1871; Baker, 1877), Gilg (1897) concluded that only a single species was involved, which he called *T. verrucosum* (Gaertn.) Oliv. This monotypic concept has been followed subsequently by most authors (Engler, 1915; Hutchinson, 1964; Brenan, 1967; Langenheim, 1973), who assumed that the natural distribution of the species was restricted to the east African coast and some adjacent islands.

Recently, in a detailed comparison of *Trachylobium* and *Hymenaea,* we presented the view that the combination of morphological, cytological (chromosome number), seedling, and chemical (resin chemistry) similarities have not resulted from convergent evolution, and thus, the two genera must be united into one, *i.e., Hymenaea* (Langenheim and Lee, 1974).

In 1837, Vogel described a new species, *Hymenaea splendida,* from northeastern Brazil. Using the basic framework set by Hayne (1827, 1830), Bentham (1870) rearranged the genus in *Flora Brasiliensis* by adding one new species (*H. eriogyne*) and two varieties (*H.*

---

*Other than sections *Hymenaea* (as *Courbaril*) and *Trachylobium,* Baillon recognized *Peltogyne* as the third section of the genus *Hymenaea.*

stigonocarpa var. *pubescens* and *H. splendida* var. *longifolia*), combining three described species with other *Hymenaea* species and transferring four (including *H. floribunda*) to the genus *Peltogyne*. Thus, he recognized six species and two varieties of *Hymenaea* in Brazil (*H. courbaril, H. stilbocarpa, H. stigonocarpa* and its variety *pubescens, H. martiana, H. splendida* and its variety *longifolia,* and *H. eriogyne*), and the Mexican *H. candolleana.*

For 50 years after Bentham, no major work was done in the genus *Hymenaea.* Several new names appeared in the literature that were later either considered to be illegitimate or were made synonyms of other extant *Hymenaea* species.

The description (Huber, 1909) of *H. oblongifolia* and *H. parvifolia* from the Amazonian rain forest is significant in understanding the taxonomic history of the genus, since we believe *H. oblongifolia* to be the "ancestral" stock of American *Hymenaea* species having the closest tie to its African counterpart, *H. (Trachylobium) verrucosa* (Langenheim and Lee, 1974).

Huber's work was followed by more extensive collecting and description by Ducke (1935, 1949), the indefatigable investigator of Brazilian legumes. In a period of 38 years, he described seven new species and two varieties of *Hymenaea (H. adenotricha, H. altissima, H. intermedia, H. palustris, H. reticulata, H. rubriflora, H. velutina,* and *H. courbaril* vars. *obtusifolia* and *subsessilis).*

In 1926 Kleinhoonte described *H. multiflora* from Surinam, but it was later (Amshoff, 1939) synonymized with the widely distributed *H. courbaril.* In 1931 Sandwith described *H. davisii* from Guyana. Then 1949 *H. torrei* was described from Cuba by León. Of interest was the fact that on the basis of the fruit character of this species, León suggested that *Trachylobium* might be made a section of *Hymenaea.*

No additional species of *Hymenaea* were described until 1971 when Paes described *H. travasii* from Colombia. In 1973 Lee and Langenheim described a new species, *H. aurea,* and a new variety (*H. oblongifolia* var. *latifolia*) from southern coastal Bahia. Subsequently, *H. maranhensis* and *H. courbaril* var. *villosa* and *H. rubriflora* var. *glabra* were described by Lee and Langenheim (1974a). Several nomenclatural changes were also made in these papers. The genus *Hymenaea* as circumscribed in this monograph consists of two sections and fourteen species including their varieties.

TRIBAL DELIMITATIONS OF DETARIEAE (CYNOMETREAE), AMHERSTIEAE AND SCLEROLOBIEAE

The genus *Hymenaea* was originally assigned to the tribe Cassieae by de Candolle (1825a, b) under Suborder Caesalpinioideae of Leguminosae (Fabaceae). It was placed in tribe Amherstieae by Bentham (1840), and only recently was considered to belong to Cynometreae Benth. emend. Léonard (Langenheim, 1973). Heywood (1971) agreed to placing *Hymenaea* in the emended Cynometreae but pointed out the priority of the name Detarieae.

The shift in tribal position of *Hymenaea* reflects the long controversial separation of tribes Amherstieae and Cynometreae, both established by Bentham in 1840. The position of the stipe of the ovary relative to the wall of the receptacle (Bentham, 1840; Taubert, 1891; Burkart, 1952) and the number of ovules (Bentham and Hooker, 1865; Harms, 1915) have been used as key characters for the separation of these tribes. Because of the increased number of genera (30 in Bentham's original work as contrasted with 87 in that of Heywood) and, more importantly, because the delimitations used previously were often minute and overlapping, the two tribes have become less distinct. For this reason, Harms (1915) questioned the separation of the two tribes, and Baker (1926) actually united them under Amherstieae. Dwyer (1954) went one step further by combining Cynometreae and the exclusively New World Sclerolobieae into Amherstieae. This union of tribes, however, does not provide a generally satisfactory account of their natural relationships since they are still considered by many systematists to be distinct groups (Fasbender, 1959; Turner and Fearing, 1959; Knaap-Van Meeuwen, 1970; Irwin, pers. comm.). This unified concept was, however, adopted in the twelfth edition of the *Syllabus der Pflanzenfamilien* (Melchior, 1964). The picture remained confused until Léonard (1957), who thoroughly revised the African Amherstieae-Cynometreae complex, singled out one character, *i.e.*, whether or not the bracteoles are valvate during anthesis, for separating the two tribes. This treatment was acclaimed by Brenan (1967) as "brilliant." The remarkable consistency of this character in tropical African genera has led to a recent reconsideration of the tribal position of *Hymenaea.* We agree that Léonard's analysis of the tribes is probably the best for evolutionary considerations. By Léonard's interpretation all Old and New World leguminous, resin-producing genera belong to Cynometreae, whereas all these genera are included in Amherstieae according to Bentham's concept (Langenheim, 1973). Heywood (1971) in his worldwide review of tribes of Caesalpinioideae agreed with Léonard's view, but revived Detarieae DC. (de Candolle, 1825b) as the older name.

Others, however, are not in agreement with this view. Hutchinson (1964), using the same characters as Léonard (1957), reached an entirely different taxonomic treatment. He noted that valvate bracteoles have arisen in those flowers in which reduction of the calyx is complete or almost complete. He thus regarded enlargement of bracteoles as compensation for loss of a protective envelope and noted that in several genera these have become thick and valvate, simulating an ordinary calyx to a marked degree. Hutchinson regarded such genera as the most highly evolved group of Caesalpiniaceae (Caesalpinioideae) and treated them as a new subfamily, Brachystegioideae. Brachystegioideae essentially coincides with Léonard's revised Amherstieae. Apart from Brachystegioideae, Hutchinson divided the remaining genera on the basis of the leaf arrangement into five highly artificial groups. *Hymenaea* and other resin-producing genera fall nearly equally into his Groups II and IV. His lack of consideration of certain floral characters, which are usually considered to be of evolutionary importance, led to assembly of such genera as *Hymenaea, Cassia,* and *Bauhinia* in the same group. Thus we share Heywood's criticism:

> One disadvantage of Hutchinson's treatment of Caesalpinioideae is that we still do not possess a comprehensive, reasonably acceptable ('Natural') classification of the subfamily from which the position of all genera can be readily determined. (Heywood, 1971: 12-13)

In a detailed study of the pollen of the American Amherstieae-Cynometreae-Sclerolobieae complex, Fasbender (1959) regarded the three tribes as discrete units. Thus, pollen characters seem to support Bentham's original division of tribes. Unfortunately, an effective comparison of Fasbender's work with that of Léonard's cannot be made, as they put their emphases on genera occurring on different continents. However, Fasbender pointed out that pollen characters do not seem to support Léonard's tribal revision.

Kooiman (1960), summarizing the results obtained by means of the "amyloid-reaction" on members of Leguminosae-Caesalpinioideae, concluded that amyloid-containing genera are found exclusively in Amherstieae, Cynometreae, and Sclerolobieae (all sensu Bentham), although amyloids are strangely lacking in the genera *Colophospermum, Goss-*

weilerodendron, and *Oxystigma.* His findings seem to favor maintenance of one large tribe Amherstieae, as suggested by Dwyer (1954). In a revision of Indo-Malayan and Pacific Cynometreae, Knaap-Van Meeuwan (1970) preferred Taubert's (1891) treatment, *i.e.,* essentially Bentham's (1840) classification, although she noted that the six genera listed in her study all have nonvalvate bracteoles, a character used by Léonard (1957) as the major delimitation of African Cynometreae.

In summary, Léonard's (1957) tribal delimitation and Heywood's (1971) substitution of de Candolle's names are adopted in this study, although delimitation of tribes Detarieae and Amherstieae is by no means completely settled. This uncertainty probably results from lack of work on these tribes on a cosmopolitan basis. They are still poorly studied in the New World and are least known in Asia. Only when studies of this kind are completed, will we have a better understanding of the two tribes and their evolutionary history.

GENERIC RELATIONSHIPS

Of the fifty-one genera listed under tribe Detarieae by Heywood (1971), our discussion will be limited to relationships of *Hymenaea* with other resin-producing genera only. Langenheim (1973) has presented an analysis of systematic relationships, distribution patterns, resin chemistry, and secretory mechanisms of these resin-producing genera, indicating the close affinities between those occurring in South America and Africa. Even among these apparently closely related genera, too little evidence is available to establish clear phylogenetic affinities between them. It would be even more difficult to attempt any meaningful interpretation of the scattered information presently available on the other genera of tribe Detarieae.

Except for *Trachylobium,* which is here considered to be congeneric with *Hymenaea* (Langenheim and Lee, 1974), the New World *Peltogyne* has usually been considered or inferred to be the genus most closely related to *Hymenaea* (Ducke, 1935, 1938; Hutchinson, 1964; Martinez, 1960). In fact, four species of *Peltogyne* were originally described in *Hymenaea,* and Baillon (1870) included *Peltogyne* as a section of *Hymenaea* in his *Histoire des Plantes.* The confusion is understandable: both genera are bifoliolate, have flowers with an elongated hypanthium, 4 sepals, 5 petals, 10 stamens, and contain resinous pockets in their leaves, petioles, young stems, and floral parts, although no mass resin production from secondary tissue has been observed in *Peltogyne* (in contrast to *Hymenaea*). When only vegetative parts are available, separation of the two genera is difficult; evidence from fruit, foliar, and seedling characters, however, suggests a more distant relationship.

The fruit of *Peltogyne* is dehiscent, subtriangular to obliquely orbicular, flat, without pulpy endocarp, and usually one-seeded and thus may be distinguished from the fruit of *Hymenaea,* which is indehiscent, ovoid to oblong, and with pulpy endocarp surrounding the seed or seeds.

The leaf of *Peltogyne* is bifoliolate; however, in most species the leaflet bears a distinct petiolule (Fig. 1a). The length of the petiolule in some species, such as *P. catingae,* can be up to 8 mm. The two halves of most *Peltogyne* leaflets are sometimes symmetrical with more or less corresponding secondary veins on both sides. In *Hymenaea,* the petiolule is twisted, bare on one side, and partly covered by the lamina of the leaflet on the other (Fig. 1a). We consider this kind of petiolule to be significantly different from petiolules in *Pelotgyne* and most "typical" leguminous genera. Dr. J. A. Wolfe of the U. S. Geologi-

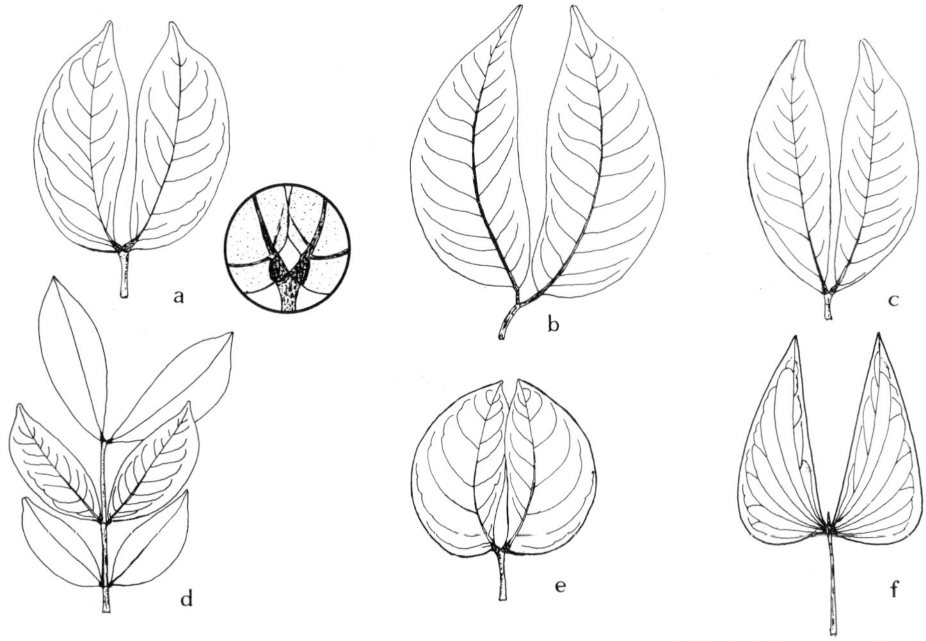

1. Leaves. a. *Hymenaea courbaril* (inset showing detailed structure of petiolule). b. *Peltogyne paniculata.* c. *Cynometra elmeri.* d. *C. malaccensis.* e. *Guibourtia schliebenii.* f. *Colophospermum mopane.*

cal Survey agrees with this view and notes that this type of petiolule is unusual in members of the Leguminosae. The two halves of the *Hymenaea* leaflet are highly unequal, and several secondary veins are usually missing from the inner half. This kind of leaf could have been derived from reduction of a multijugate impari- or paripinnate strand having petiolular structure similar to that commonly found in the living genus *Cynometra* (Fig. 1c, d). Thus, the bifoliolate leaves found in *Hymenaea* and *Peltogyne* may have resulted from convergent evolution.

Both fruit structures and similarities of immature inflorescences in some species of *Peltogyne* with those in *Copaifera* suggest their affinity. Dwyer (1958), on the other hand, suggested that *Peltogyne* is more closely related to *Cynometra* than to any other genus.

Additional evidence against an immediate affinity of *Hymenaea* and *Peltogyne* comes from the seedling characters. These two genera clearly show distinct seedling types (Fig. 2a, b), thus possibly indicating a relatively remote relationship between them. The importance of blastogenic characters in distinguishing genera within the tribe Cynometreae (Detarieae) has been emphasized by Léonard (1957).

The African *Guibourtia* has bifoliolate leaves (Fig. 1e), and it shares both a similar resin secretory system and type of seedling with *Hymenaea* (Fig. 2c-e). However, *Guibourtia* has dehiscent fruit and apetalous flowers that are more similar to *Copaifera* than *Hymenaea.*

Recently, Stein (1975), in a study of foliar and stipular development in *Hymenaea courbaril* vars. *courbaril* and *stilbocarpa* noted that there are three leaflet primordia in the very young bud. One of these primordia soon degenerates and the remaining two develop to form the two leaflets that characterize the entire genus. In the monotypic African *Colophospermum,* in addition to the two "normal" leaflets, a much smaller and appar-

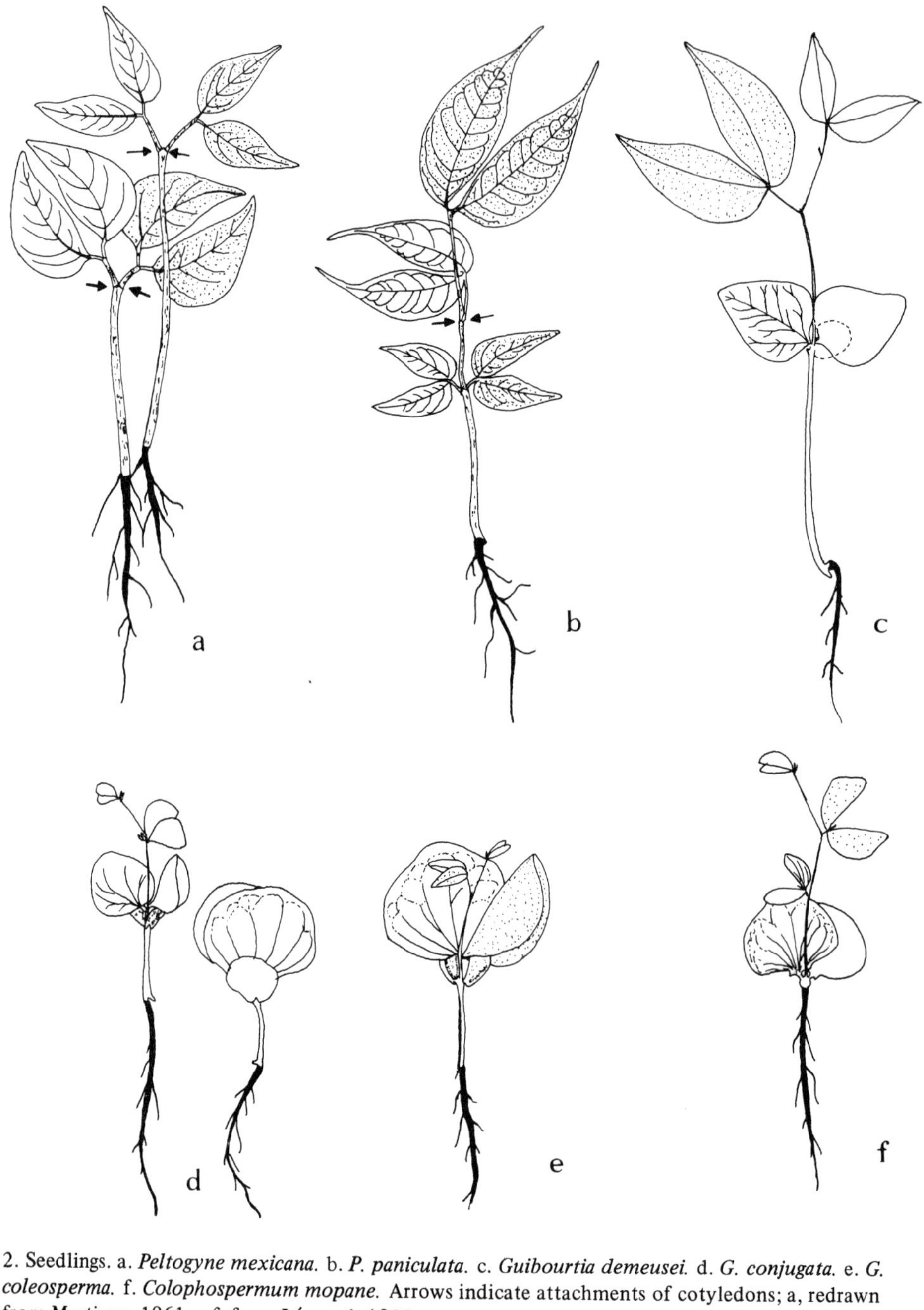

2. Seedlings. a. *Peltogyne mexicana.* b. *P. paniculata.* c. *Guibourtia demeusei.* d. *G. conjugata.* e. *G. coleosperma.* f. *Colophospermum mopane.* Arrows indicate attachments of cotyledons; a, redrawn from Martinez, 1961; c-f, from Léonard, 1957.

ently vestigial leaflet is also present (Fig. 1f). Whether or not the presence of such a "third leaflet" in the very early developmental stage (as in *Hymenaea*) or throughout the life span of the leaf (as in *Colophospermum*) reflects a strong relationship between the two genera remains to be seen. Evidence derived from seedling characters, however, would clearly dispute any immediate relationship between the two as the seedling of *Colophospermum* (Fig. 2f) is characterized by a subsessile hypocotyl and two foliaceous cotyledons (Léonard, 1957), which results in a significantly different seedling type from that in *Hymenaea* (Figs. 4, 5).

Similarities in floral characters (cf. Vogel, 1968, Fig. 13) in *Hymenaea* and the neotropical genus *Eperua* also suggest a possibly close relationship. *Eperua,* however, differs significantly from *Hymenaea* in: (1) the often oblique, smooth, plano-compressed, two-valved fruits, (2) paripinnate leaves, (3) the single, large, and very broad petal which is never punctate,* and (4) the filaments of at least nine of the stamens are united basally in a short, equilateral to zygomorphic tube.

When all the resin-producing genera are considered (Table 1), it is evident that in contrast to the predominantly dehiscent and pulpless fruits within this group, *Hymenaea* is the only genus characterized by woody and indehiscent fruits with seeds surrounded by pulpy endocarp. This unique fruit type in *Hymenaea* presents some difficulties in assessing the immediate affinity of the genus to other genera. However, the foliar characters, particularly the paripinnate leaflets and the petiolular structure, and the presence of woody and indehiscent fruit among some *Cynometra* species indicate a great similarity to those in species of *Hymenaea.* These characters support the assumption that *Hymenaea* could have been derived from this group. The pantropical distribution of *Cynometra* [thirty species in the Neotropics (Dwyer, 1958), twenty-five in tropical Africa (Léonard, 1951), and twenty in the Asiatic tropics (Knaap-Van Meeuwen, 1970)], together with other evidence mentioned above suggest that this genus might have also served as the parental stock of other genera such as *Guibourtia, Peltogyne* and *Copaifera* (Langenheim, 1973; Langenheim and Lee, 1974).

COMPARATIVE MORPHOLOGY AND GENERAL ANATOMY

HABIT

Hymenaea consists of usually large but sometimes small or medium-sized trees, which are primarily evergreen. However, some are facultatively deciduous and one is known to be obligately deciduous. The genus includes emergent trees, reaching a height of more than 25 meters and a girth of more than 1 meter, in tropical rain forests and small, shrubby trees in grassy areas or in thorn forests. In tropical rain forest or in humid evergreen forest along the Atlantic coast, the trees are usually characterized by a straight, cylindrical trunk with a spreading, flat-topped crown. In more open ecosystems they are much shorter, with a well-developed umbrella- or cone-shaped crown. The growth form within a single species varies considerably with the habitat. For example, *H. courbaril* and *H. parvifolia* are emergent trees when growing in the forest; they become shrubby trees when growing in either grassland or highly disturbed areas.

Most species have a smooth, silver-grey bark, usually covered with scattered lichen thalli. The bark can be more than 3 cm. thick in some of the *hylaea* species and was used

---

*Information kindly provided by Dr. R. S. Cowan of Smithsonian Institution.

TABLE 1. Comparison of Morphological Characters among the Resin-Producing Genera

			Gossweilero-dendron	Colophos-permum	Tessmlannia	Oxystigma	Daniellia
Distribution			Trop Afr (1)	Trop Afr (1)	Trop Afr (11)	Trop Afr (5)	Trop Afr (12)
Resin-secreting System			Pockets?	Pockets	Pockets?	Pockets?	Canals
LEAVES	Stipules		2, Caducous	2, Caducous	2, Caducous	2, Caducous	2, Inter-petiolar
LEAVES	Leaflets	Arrangement	Imparipinnate	Opposite	Paripinnate	Pari- or imparipinnate	Paripinnate
LEAVES	Leaflets	Number	6-10	2	Several pairs	1-9	1 to few pairs
LEAVES	Leaflets	Pullucid-punctate	+	+	+	+	+
LEAVES	Leaflets	Petiolule	Twisted	Twisted	Twisted	Not twisted	Twisted
Inflorescence			Panicle (raceme)	Panicle or raceme	Panicle or raceme	Panicle (raceme)	Panicle or raceme
FLORAL PARTS	Hypanthium		Short	Short	Short	Short	Elongate
FLORAL PARTS	Disc		+?	+?	+	+	+
FLORAL PARTS	Mass of nectar tissue		-	-	-	-	?
FLORAL PARTS	Calyx lobes		4 (-5)	4	4 (-5)?	5 (-6)	4
FLORAL PARTS	Petals		0	0	5	0	5 (-1)
FLORAL PARTS	Stamens		10 (-8)	20-25	10 (1+9)	10 (8)	10 (1+9)
FLORAL PARTS	Stipe		Excentric	Excentric	Excentric	Excentric	Excentric
FRUIT	Dehiscent		-	-	+	-	+
FRUIT	Surface		Veined	Smooth	Smooth	Veined	Faintly veined
FRUIT	Shape		Samaroid	Raniform, flat	Oblong to ovoid compressed	Oval to ellipsoid compressed	Oblong to ellipsoid flat
FRUIT	Pulp		-	-	-	-	-
FRUIT	Texture		Coriaceous	Coriaceous	Woody	Coriaceous	Stiffly papery to thinly woody
FRUIT	No. of seeds		1	1	1-4	1	1
SEED	Shape		Oval to ellipsoid	Reniform, flat	Oblong to ovoid	Oval to ellipsoid	Oval to ellipsoid
SEED	Aril		?	?	Small	?	Small
SEED	Funicle		+	?	?	?	+, very long
Seedling			Hypogeal	Epigeal	Epigeal	Hypogeal	Epigeal

TABLE 1 *(Continued)*

Guibourtia	Copaifera	Cynometra	Hymenaea	Eperua	Peltogyne	Prioria
Trop Afr (13) America (1?)	Trop Afr (4) America (25) Asia (1?)	Trop Afr (25) America (25) Asia (20)	Trop East Africa (1) Neotrop (13)	Neotrop (14)	Neotrop (22)	Neotrop (1)
Pockets	Canals	?	Pockets	Canals	Pockets	Pockets
2, Caducous	2, Caducous	2, Caducous	2, Caducous	Small or lge interpetiolar	–	Scale like
Opposite	Pari- or imparipinnate	Paripinnate	Opposite	Paripinnate	Opposite	Paripinnate
2, rarely 1	Several	1 to few pairs	2	2-6 pairs	2	2-4
+	+	?	+	+	+	+
Twisted	Not twisted	Twisted	Twisted	Not twisted	Not twisted	Not twisted
Panicle	Panicle or raceme	Panicle or raceme often fasciculate	Long- or short- panicle	Comp'd racemes	Panicle (racemes)	Panicle (spikes)
Short	Short	Short or elongate	Short or elongate	Short	Short or elongate	Short
+?	+	+	+	+	+	+
-	-	-	- or +	+	?	-
4	4	4-5	4	4	4	5
0	0	4-5	5, rarely 3 +2 petaloids	1 + 4 petaloids	5	0
12-8	10 (8)	10, rarely ∞	10	10-4 (with staminoids)	10	10
Excentric	Central	Cent., rarely excentric	Excentric	Central	Excentric	Central
+ or -	+	+ or -	-	+	+	+
Smooth	Smooth	Smooth to rugose	Smooth to verrucose	Smooth	Smooth	Veined
Oval to ellipsoid flat	Oval compressed	Ovoid to sub-veniform compressed or not	Ovoid to cylindrical compressed or not	Oblong to subquadrate plano-compressed	Orbicular to subtriangular flat	Sub-orbicular flat
-	-	- or +?	+	-	-	-
Coriaceous	Coriaceous	Coriaceous to woody	Woody	Sub-woody	Coriaceous	Sub-woody
2 (1)	1 (2)	1-2	1-many	Few-many	1	1
Oval	Oval	Ovoid to ellipsoid	Ovoid to oblong	Oval to oblong	Oval to oblong	Oval to oblong
Small	Small	Small	Small	Small	Small	Small
+	+	-	-	-	+	+
Epigeal	Epigeal	Epigeal?	Epigeal	Epigeal	Epigeal	Hypogeal

by Indians to make canoes. In *H. stigonocarpa* and particularly var. *pubescens,* the bark is very rough and characterized by deep fissures and grooves.

The branchlets are slightly flexuous, glabrous to densely covered with short hairs. They are dotted with resinous pockets and minute lenticels. The surface of the branchlets is smooth or occasionally covered by peeling epidermis when aged.

Buttress roots are common in large trees in tropical forests and in humid forests along the Brazilian Atlantic coast. Such roots are medium-sized, up to 1.5 meters tall and 1 meter wide.

LEAVES

The leaves are stipulate and alternate, characteristically punctate and bifoliolate. At the apex of the petiole, there are two points of insertion upon which the two leaflets are attached.

Stipules: Stipules occur in pairs at the base of petioles and are most frequently early deciduous. They are narrowly to broadly elliptic, obtuse, truncate at the base, and dotted with resinous pockets. The two stipules, according to Stein (1975), interlock with each other by marginal hairs and encase their leaves until they are 2-3 cm. long (or longer in some species). Then the "capsule" opens, displaying the new terminal bud at the base of a rather well-developed petiole. Thus the stipules enclose the younger structures and also sheath the leaf to which they are attached. Stein indicates that only a few species of plants share this stipular pattern. Sipules of seedlings of *H. parvifolia* usually remain attached for at least one growing season, representing a unique case in the genus. Size of stipules varies among different species and at different developmental stages. The pattern of vesture of stipules is similar to that of the leaf blades.

Petioles: All leaves in *Hymenaea* are petiolate. The petiole varies in length, rigidity, and vesture in different species and varieties. It is dotted with resinous pockets and minute lenticels.

Leaflets: The leaflets are petiolulate. The petiolule is twisted, 2-5 mm. long (the outer side is always covered by the lamina of the leaflet and is thus subsessile), dotted with resinous pockets and minute lenticels, glabrous to densely pubescent, depending on type of vesture appearing on the leaves. The lamina of the leaflet varies from ovate, obovate, or oblong to falcate, the inner half is much narrower than the outer half, coriaceous, the apex acuminate, apiculate, acute, obtuse or retuse, the base oblique, with the inner side acute to slightly rounded and the outer side slightly rounded to broadly expanded and rounded.

In form and dimensions there is great diversity among leaflets of different species and varieties. Both size and shape of leaflets have diagnostic value, particularly at the intraspecific level.

Vesture of leaflets is variable, but consistent in the same taxonomic grouping, and is employed frequently in the following treatment. The abaxial surface is usually duller in color than the adaxial.

Venation: The venation pattern according to Wolfe (pers. comm.) is *camptodromous*;* *i.e.,* the pinnately arranged secondary veins terminate at the laminar margin. Two subtypes are recognized, depending on whether or not the secondary veins are joined to-

*Terms used here are adopted from Hickey (1973).

gether in a series of prominent arches (*brochidodromous**) or are upturned, gradually diminishing apically inside the margin without forming prominent marginal loops (*eucamptodromous**). A detailed study of venation pattern of *Hymenaea* is now in progress as part of a larger study by Dr. Wolfe.

Epidermal pattern: Two types of epidermal pattern are present in *Hymenaea* (Wolfe, pers. comm.). Epidermal cells of the first type are characterized by straight walls, of the second type by sinuous walls.

INFLORESCENCE AND FLOWERS

Inflorescence: The inflorescences are of two basic types, although the differences between them are not absolute.

In the first type, the inflorescence is long-paniculate, characterized by: (1) distinct terminal and lateral racemes; (2) long, slender, and slightly flexuous branches; and (3) relatively small and numerous flowers. The second type is a short panicle when young, becoming densely corymbose when mature, and characterized by (1) branches rather short, robust, and strongly flexuous and (2) fewer and much larger flowers. Because of the nature of the branches, the terminal raceme is not apparent, while the lateral ones are relatively longer. Thus, during anthesis most flowers line up at the top, forming the rather flat surface similar to a corymb. In some populations the inflorescence is reduced such that only one to a few racemes are left. Intermediate forms indicating evolutionary trends are found between these two types of inflorescence.

Sometimes, particularly among members of sect. *Trachylobium,* the main axis of the apparent panicle bears small leaflets and the inflorescence proper thus appears to be composed of a terminal and several lateral racemose branches.

Bract and bracteoles: Bracts occur regularly at the base of inflorescence branches on the upper portion of the inflorescence. The two bracts usually enclose several young flowering buds and are early-caducous.

Flowers are short- to long-pedicellate. At the apex of each pedicel are found two bracteoles which completely envelop the very young flowering bud. They are arranged in such a way that one overlaps the other marginally, but they are never valvate, a character given by Léonard as the key delimitation for tribe Cynometreae. Bracteoles are well-developed, relatively small, ovate or orbicular, concave, pubescent outside but glabrous and reddish-brown within, dotted with resinous pockets, and usually caducous except in *H. maranhensis,* where they usually remain attached during anthesis.

Flowers: The flower consists of a hypanthium, apparently four sepals, five petals, ten stamens, and a pistil. All floral parts are dotted with resinous pockets.

The hypanthium, which is campanulate, sessile, or relatively long-stalked, is believed to result from fusion of the bases of sepals, petals, and filaments. It is very coriaceous and pubescent externally, smooth and shining within. The sepals are borne on its outer margin; petals and stamens are inserted in the inner rim. In many species, there is a prominent fleshy disc around the inner rim (Fig. 9h). Below the disc and on the inside, there is an inverted, bell-shaped mass of nectariferous tissue (Fig. 9i, j). This tissue is fleshy and differs from the surrounding wall.

Calyx: The calyx is urn-shaped, coriaceous, strongly imbricate in the bud, and usually

*Terms used here are adopted from Hickey (1973).

deciduous or occasionally persistent. The two lower sepals are fused, so that the calyx appears four-parted. Size and proportions of the calyx vary considerably among different species but are of little taxonomic value. Types of vesture appearing on both sides of the calyx provide consistent characters and are used frequently in the following treatment.

Corolla: The five petals are subequal, usually creamy white, rarely red. Shape of the petals varies from lanceolate or spatulate to ovate and obovate, and from distinctly clawed (claws to 8 mm. long) to almost sessile. Petals of all species are glabrous, except in one case where they are densely white-pilose on the adaxial surface. The petals are characterized by a prominent central midrib and several parallel, lateral veins. They are usually early-deciduous.

Androecium: The ten stamens are all fertile and arranged, more or less clearly, in two rows. The filaments are nearly the same length as the petals. In bud the filaments are folded so that the dorsifixed anthers are pressed against each other, with the anthers surrounding the pistil. During anthesis, the upper part of the folded filaments extends and exserts the anthers. The anthers are bilocular, linear, crenulated on both ends, and versatile. The stamens are early-deciduous.

Pollen: The pollen grains, according to Whittingham (unpubl. ms.) are spherical to subprolate and tricolporate. The three meridional furrows, each with a pore at the equator, extend nearly to the poles, or they may be syncolpate. The exine is punctitegillate, and in surface view with Scanning Electron Microscope, appears essentially as an arrangement of ridges, mounds, or rugae, separated by punctate depressions or grooves. The puncta in some species are partially occluded, either by a thickening of the tectum at the poles, or by the development of superposed rugae or verrucae.

Included by this generalized description are a number of distinctive exine patterns, characteristic of individual species. One feature is perhaps worthy of special mention here due to its possible significance relative to the taxonomic system. This is the minute nature of the puncta found in two species, *Hymenaea verrucosa* and *H. oblongifolia,* both of sect. *Trachylobium.* Especially when viewed with SEM, the well-developed system of ridges in these species is more evident than the obscure perforations in between. By contrast, in section *Hymenaea,* all species are conspicuously punctate, regardless of whatever other features they may exhibit.

More extensive studies of pollen of various *Hymenaea* species are now in progress (Whittingham, unpubl. data), using both light and scanning electron microscopy.

Gynoecium: The single pistil is characterized by a short or long stipe (the gynophore), which is distinctly adnate laterally to the wall of the hypanthium. The ovary is obliquely oblong, or rhomboidal and compressed, and contains two to eighteen or more ovules. The ovary provides important characters in type and distribution of pubescence, both very stable characters in this genus and hence often used in delimiting species. The ovary may be pubescent throughout, only partially pubescent, or entirely glabrous. In most apparently glabrous ovaries, however, hairs do occur at the base or along the margin, an indication that a glabrous ovary is a derived condition. The style is glabrous, emerges laterally from the ovary, is threadlike, and is as long as the stamens. The upper part of the style is usually bent and coiled inward in bud. In anthesis, the pistil uncoils and stretches with the capitate and bilobed stigma pointing upward.

Fruit and Seeds

Fruit: The fruit represents the most distinctive feature of the genus *Hymenaea* within tribe Detarieae. It is woody, indehiscent, green when young and turning light to dark brown when mature. The surface of the fruit varies from highly verrucose to smooth, but is always dotted with numerous, minute, cream white lenticels. Shape of fruit ranges from oblong to ovate to obovate and, in cross-section, from spherical or terete to strongly compressed. In some species, the fruit is usually ovate to obovate and one- to few-seeded, and in others it is usually large, oblong, occasionally rhomboidal, and many-seeded. A suture is prominent in some species but absent in others. The seed or seeds are surrounded by a dry, floury-filamentous, yellowish brown pulp said to be endocarpic (Nazif, 1914), but whose origin is still subject to controversy. Pascher and Pohl in their article on "Fortpflanzung" in the *Handbuch der Naturwissenschaftern* (*cf.* Van der Pijl, 1956b) defined the pulp as an endocarp product. Burkart (1952), however, characterized fruit of *Hymenaea* as having a thick and crustaceous epicarp, a floury and abundant mesocarp, and a thin and articulate endocarp. Such endocarp has not been found among fruits of all the species we have examined. Therefore we accept Nazif's interpretation that the pericarp of *Hymenaea* has a thin, usually glossy epicarp, a thick, rather woody mesocarp, and a floury endocarp. The pulpy endocarp is edible, sweet, with a strong odor, and is eaten by local residents, especially children, and by such animals as wild pigs and some rodents.

Seeds: The seeds are ovoid to ellipsoid, compressed or not, and with a bony, light brown to dark brown, shining and smooth testa. A small hilum-lump is located at the base of the seed, with the aril minute or absent.

The cotyledons are fleshy, creamy white, and separated at the circumference by a fissure. A dorsal median groove is found in several species but is absent in the others. Thus, intrusion of the testa into the cotyledons produces two or four shallow furrows.

In summary, among the morphological characters the two basic types of inflorescence are particularly useful in distinguishing the two sections of *Hymenaea.* Sect. *Trachylobium* is characterized by a long-paniculate inflorescence and sect. *Hymenaea* by a corymbose inflorescence, although the differences between them are not absolute. Several other characters, such as shape of petals, vesture of ovary, shape and size of fruit, and number of seeds also have diagnostic value in separating the two sections (Fig. 3).

SEEDLINGS

The significance of seedling (blastogenic) characters in understanding relationships among the African Detarieae-Amherstieae has been demostrated by Léonard (1957). In his comparative study of seedlings with respect to systematic revision of the African genera, he emphasized that the seedlings of all the species of one "good" genus have the same type of structure and there should be a clear correlation between adult morphological and seedling characters. He was convinced of the greater relative importance of adult morphological characters, but within this framework proposed four guidelines for use of seedling characters in a systematic study.

A thorough study of *Hymenaea* seedlings, including twelve of the fourteen extant species, was inspired by Léonard's work. The purpose of this investigation was (1) to test

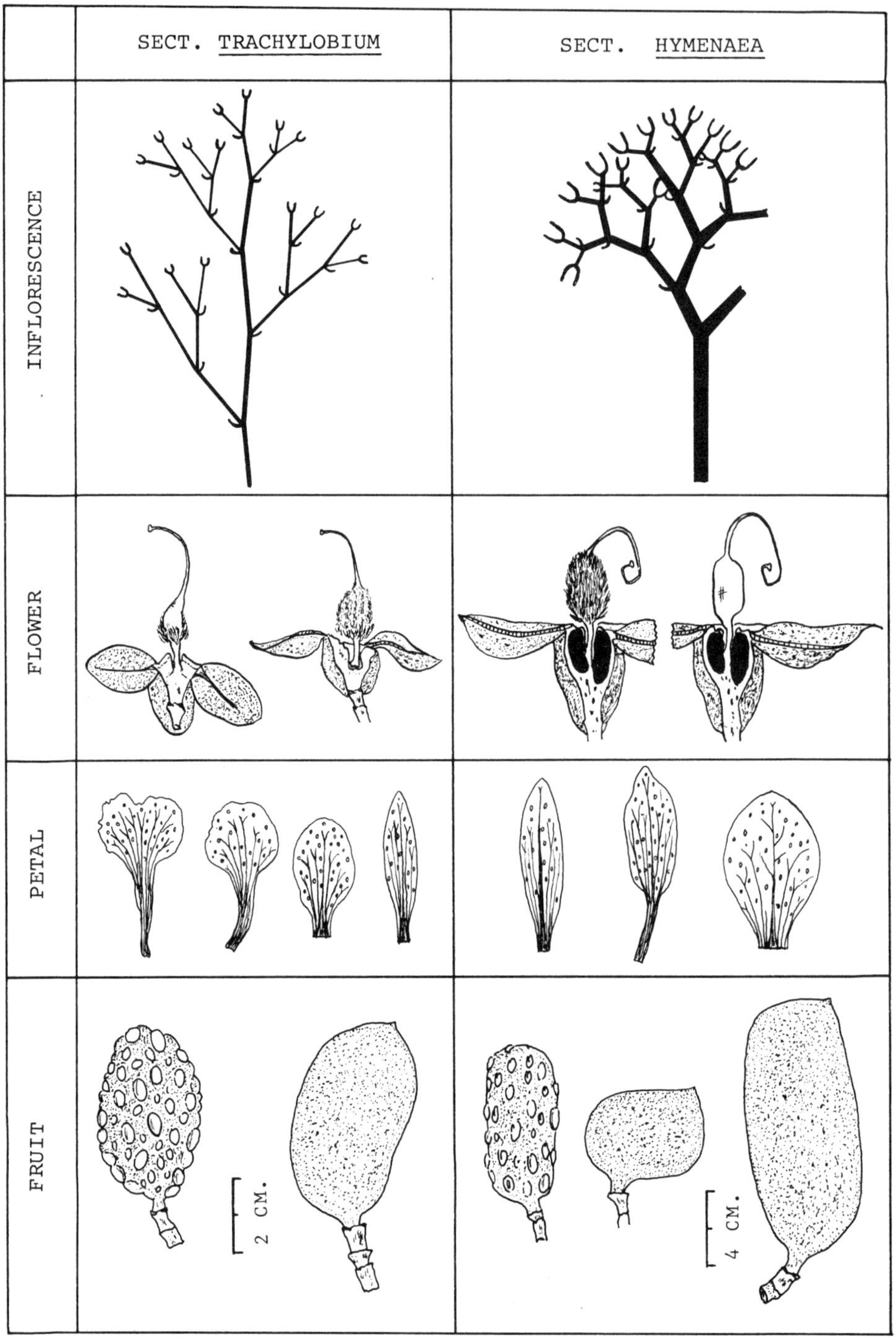

3. Comparison of inflorescence, flower, and fruit characters of the two sections of *Hymenaea.*

the homogeneity of the genus *Hymenaea* and (2) to give a detailed account of the seedling characters of *Hymenaea* which will ultimately be useful for a comparative study at the tribal level in the New World.

Seeds of various *Hymenaea* species were collected in the field by us or were sent to us by local botanists in Africa. Seeds were first scarified by piercing a small opening in the testa along the edge, and then were moistened and coated with a thin layer of fungicide (Thiram, 75% Bis-dimethyl-thiocarbomyl diasulfide). After pretreatment, they were sown in groups according to species in plastic pots with sand, and put into a growth chamber with continuous light (day temp. 35° C, night 28° C, relative humidity 55%). Germination took place within 8 to 18 days. After the first leaf reached full growth and the subsequent leaves had begun to emerge, the young seedlings were transplanted to clay pots with specially mixed soil and placed in the greenhouse with natural lighting (day temp. 28° C, night 24° C, relative humidity 60-75%).

At germination the hypocotyl emerges and the fleshy cotyledons are pushed above the ground. The root begins to develop rapidly. At this point the hypocotyl starts to elongate and straighten while the epicotyl (if apparent) has developed as has the primary leaf. There are some differences in length of time necessary for development of the subsequent stem and leaves. *Hymenaea courbaril* and *H. martiana* are usually the fastest-growing species and *H. parvifolia* the slowest.

Observations were made for about two months on each set of seedlings. In those for which no seed was collected, information was gathered from seedlings collected in the proximity of a known parent tree.

The seedlings of the twelve species of *Hymenaea* (including thirteen varieties) investigated are of a single basic type. They are epigeal with a primary leaf composed of two leaflets, differing in form from the subsequent alternate bifoliolate leaves. In addition to basic seedling type, the presence of a distinct collar between the base of the hypocotyl and the root near soil level as well as a median dorsal groove in the cotyledon have been considered by Léonard (1957) to be of value in distinguishing genera of the African "Cynometreae-Amherstieae" complex. The collar is found in all seedlings of *Hymenaea* taxa thus far investigated. In *H. (Trachylobium) verrucosa,* the testa intrudes into the two grooves and into the fissure which separates the two cotyledons, forming four shallow furrows. This character was used incorrectly by León (1949) to distinguish the genera *Hymenaea* and *Trachylobium.* The dorsal grooves are clearly visible in the young seedlings after the cotyledons emerge and the testa is shed. Dorsal grooves similar to those in *H. verrucosa* are found also in *H. oblongifolia* var. *latifolia, H. parvifolia,* and *H. reticulata,* all presumed to be among the more primitive New World taxa. In the remaining species, in which the dorsal median groove is not prominent, a vestigial indentation is always evident (Fig. 4a-c). This indicates the constancy of this character throughout the genus. Thus, these seedling characters are in agreement with the morphological characters discussed previously in supporting the homogenity of *Hymenaea.*

However, within this basic scheme, two subtypes can be recognized. In subtype I, the epicotyl is short or absent and the primary leaf occurs at or very close to the attachment of the cotyledons. In subtype II, the epicotyl is relatively long and the separation of the primary leaf and the attachment of the cotyledons is distinct. A diagrammatic representation of the subtypes and groupings of *Hymenaea* species within them is given in Fig. 5.

The two subtypes of seedlings represent responses to two ecologically distinct conditions (Table 3), but the possible significance of epicotyl length in terms of habitat dif-

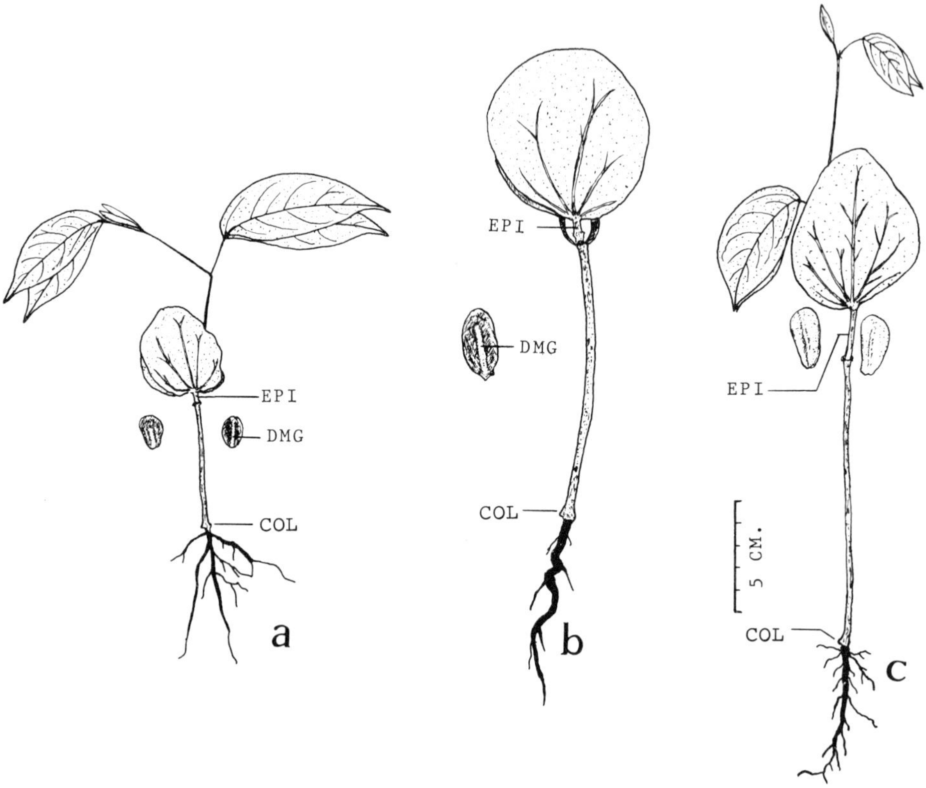

4. Seedlings of *Hymenaea* species. a. *H. verrucosa.* b. *H. oblongifolia* var. *latifolia.* c. *H. courbaril* var. *courbaril.* (EPI = epicotyl; COL = collar; DMG = dorsal median groove.)

ferentiation is not clear. It is evident that species belonging to subtype I occupy wetter habitats whereas those in subtype II are found in drier ones. The two ecological groupings, when correlated with other characters, are useful for a better understanding of the evolutionary history of the genus. They do not, however, concur with the two sections recognized taxonomically.

Furthermore, dimension and shape of leaves in seedlings and saplings often differ considerably from those in adult trees, being characterized by much larger leaflets and longer drip tips. In some extreme cases, such as those in *H. oblongifolia* vars. *oblongifolia* and *palustris* and *H. rubriflora* var. *rubriflora,* seedling leaves differ from those of adult trees in texture, shape, size, and vesture pattern.

Dimorphism during different developmental stages may indicate an adaptive trend for the rain-forest species, because leaves on emergent branches differ significantly from those of seedlings or saplings which grow on the poorly illuminated forest floor. Thus, being much larger, thinner, and devoid of pubescence in the examples mentioned, the leaves may provide a better adaptation to the densely shaded habitat and enhance the chances of survival for younger plants.

In summary, a description of the seedlings of *Hymenaea* is given below:

Seedling with epigeal cotyledons. Collar embellished at junction of hypocotyl and

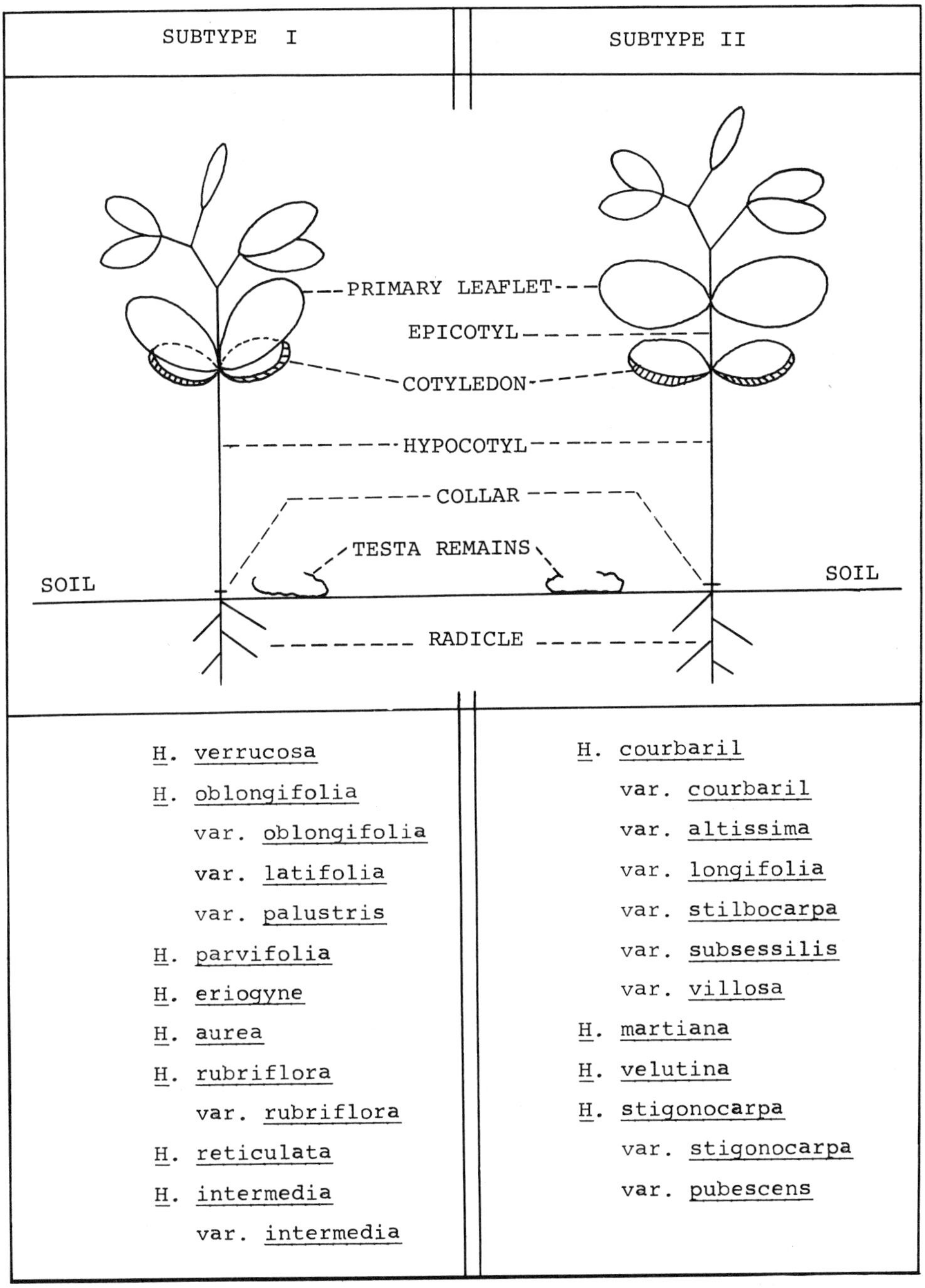

5. Diagrammatic representation of seedling subtypes with grouping of *Hymenaea* species within them.

young root with a large, white disc which atrophies with age (or, occasionally, the collar not evident). Hypocotyl 2.5-25 (+) cm. long, densely golden-puberulent when young, ± glabrate (or glabrescent). Cotyledons clearly hollowed out at base, plump, thick, and smooth, either transversely indented by a clear median groove on dorsal side, or not, the separation between the two cotyledons marked on each side by a deep groove. The two cotyledons of young seedlings thus form an ensemble provided with 4 or 2 characteristic furrows (in seeds such furrows are filled with intrustions of the testa, see previous section). Epicotyl absent or up to 4 cm. long, dotted with resinous pockets and lenticels, usually glabrous. Primary leaf unijugate, with a rachis represented by a small, pointed, tardily deciduous structure or such structure not evident. The two primary leaflets subsessile, appearing like two opposite leaves, suborbicular to cordate, rounded or obtuse, of different form than leaflets of the subsequent leaves, palmately veined with 3 to 5 major veins, glabrous or occasionally thinly golden brown tomentose, densely dotted with resinous pockets. The new bud emerges from junction of the two primary leaflets; primary internode glabrous to densely tomentose, the following leaves stipulate, alternate, distinctly petiolate and bifoliolate. These leaves either identical with (but often larger and with longer drip tips) or entirely different from leaves of adult tree, with densely translucent punctae and either brochidodromously or eucamptodromously veined. Epidermis of hypocotyl is shed within two months but usually remains attached for some time.

CHROMOSOME COUNTS

Atchison (1951) reported the first chromosome counts of any species of *Hymenaea,* indicating the diploid number to be 2n = 24 for *H. courbaril* and 2n = 16 for *H. verrucosa* (as *Trachylobium verrucosum*). The chromosome number of *H. courbaril* was later confirmed by Berger *et al.* (1958) and by Bawa (1973). It was not until recently that the chromosome number of *H. verrucosa,* reported by Atchison from a tree growing in the Atkins Garden in Cienfuegos, Cuba, was challenged. In a detailed comparative study of *Hymenaea* and *Trachylobium,* which resulted in reunification of the two genera (Langenheim and Lee, 1974), we felt the need to check the published count of *H. verrucosa* and the chromosome number of other *Hymenaea* species. Atchison's count of *H. verrucosa* could not be confirmed.

All chromosome counts were made from root tips pretreated with a saturated aqueous solution of paradichlorobenzene for 2 to 4 hours at room temperature and then fixed in Carnoy's fixative (100% EtOH-glacial acetic acid, 3:1 v/v). Squash preparations of chromosomes were made by methods suggested by Whittingham (pers. comm.). Both leuco-basic fuchsin solution and 0.25% orcein in 45% acetic acid were employed in the named sequence. Hydrolysis was necessary to facilitate better cell spreading and was done prior to the fuchsin treatment, in 1N HCl at 60° C, for 8 to 12 minutes, depending on thickness of root tips.

Seeds used in this study were collected either directly from the tree or picked up from the ground in the immediate vicinity of a known tree, and whenever possible herbarium specimens were prepared at the same time. All the voucher specimens were deposited in the University of California Herbarium, Berkeley (UC).

The chromosome number of somatic cells in two populations of *H. verrucosa* was first reported by Whittingham (*cf.* Langenheim and Lee, 1974) to be 2n = 24. These counts were subsequently confirmed in this study. Thus, the chromosome number of *H. verrucosa* is believed to be 2n = 24, which is the same as for the other eleven species and nine varieties (Lee and Langenheim, 1974b). A summary of all the reported chromosome counts is given in Table 2. Selected camera lucida drawings of mitotic chromosomes of various taxa of *Hymenaea* are shown in Fig. 6.

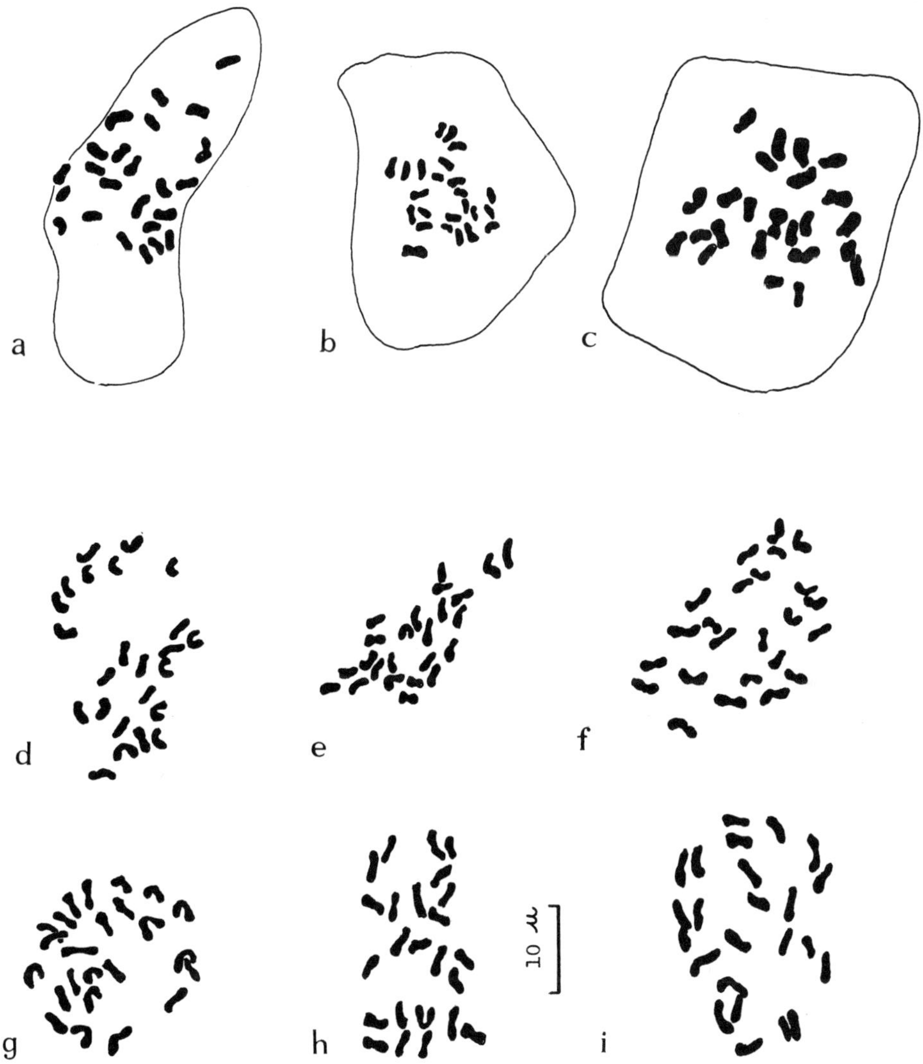

6. Camera lucida drawings of somatic chromosome complement of various *Hymenaea* species. a. *H. velutina* (YTL 115). b. *H. courbaril* var. *courbaril* (JHL 5743). c. *H. verrucosa* (JHL 5587). d. *H. parvifolia* (JHL 5618). e. *H. courbaril* var. *altissima* (JHL 5643). f. *H. courbaril* var. *stilbocarpa* (JHL 5642). g. *H. aurea* (JHL 5644). h. *H. stigonocarpa* var. *pubescens* (JHL 5763). i. *H. eriogyne* (YTL 116).

TABLE 2.
Reported Chromosome Counts in *Hymenaea*
(Data derived from Lee & Langenheim, 1974b unless otherwise indicated.)

Species	Chromosome no. 2N	Remarks
	16	Atchinson, 1951
H. verrucosa	24	Langenheim & Lee, 1974
	24	
H. oblongifolia		
var. *oblongifolia*	24	
var. *latifolia*	24	
H. parvifolia	24	
H. eriogyne	24	
H. aurea	24	
H. rubriflora		
var. *rubriflora*	24	
H. reticulata	24	
H. courbaril		
	24	Atchison, 1951
var. *courbaril*	24	Berger, *et al.* 1958
	24	Bawa, 1973
	24	
var. *altissima*	24	
var. *stilbocarpa*	24	
var. *subsessilis*	24	
var. *villosa*	24	
H. martiana	24	
H. velutina	24	
H. stigonocarpa	24	
var. *pubescens*	24	

RESIN CHEMISTRY

Resin Secretory Systems and Chemistry Among Genera in the Detarieae

All of the known resin-secreting leguminous genera belong to the subfamily Caesalpinioideae and the tribe Detarieae (Langenheim, 1973). All of the genera (except possibly *Cynometra*) previously discussed regarding their relationships to *Hymenaea* produce resins. Therefore, resin is a chemical character worthy of consideration at the generic as well as the specific or population level.

Resin production is particularly interesting in the Detarieae because of the variety of anatomical arrangements for secretion within different organs or at different stages of maturity in the same organ in the same plant (Guignard, 1892; Moens, 1955; Langeheim, 1967, 1969). Different genera or groups of genera are also characterized by different secretory systems.

Primarily because of their commercial importance, some of the Detarieae resins, commonly called "copals," have been analyzed chemically. Various physical and chemical properties, such as melting point, solubility, and saponification values, were commonly used to characterize resins until relatively recently; now more modern techniques, such as gas chromatography, mass spectrometry, and infrared spectrometry, are available to facilitate structural analysis. Unfortunately, most of the earlier analyses did not provide useful data for botanical purposes. Extensive investigations of confierous resins, emphasizing compositional differences, have shown that these resins can be useful taxonomically (*e.g.*, Zavarin and Frank, 1965; Zavarin *et al.*, 1966; Mirov, 1967; Zavarin and Snajberk, 1973; Adams and Turner, 1970). These studies also demonstrated that resin composition may vary greatly in tissues of different organs, such as leaf, wood, bark, or fruit, and they must be considered separately in any chemosystematic interpretation.

The first work on leguminous resins that could be used in any chemosystematic context was based upon infrared spectra of Congo copals. Hellinckx (1955) demonstrated that infrared spectra of carbon tetrachloride extracts of copals were useful in determining the botanic source of various species of *Guibourtia, Copaifera, Daniellia,* and *Tessmannia.*

Infrared spectra made from pellets including the untreated resin have also been used in identifying the sources of some fossil resins (Langenheim and Beck, 1965, 1968; Langenheim, 1968, 1969, 1972). In the polymerization of resins in fossilization, all simple functional groups are preserved except carbon-carbon double bonds, whereas skeletal frequencies are damped but not extinguished. Only major constituents of the complex of terpenoids comprising resins would be expected to give strong absorption bands and, although absolute identity of spectra can never be expected, the presence of similarities indicates structural similarity of the major constituents. The close relationship in spectra between recent and fossil resins indicates that at least the skeletal structure of the major constituents of some taxa appears to be relatively stable over millions of years. In the course of determining that amber from Mexico, Colombia, and Brazil was derived from *Hymenaea* and amber from Angola came from *Copaifera,* spectra were made of resin from numerous species of *Hymenaea, Copaifera, Guibourtia* and *Daniellia.* Characteristic spectra appear to be representative of either genera or species complexes in the various groups studied (Langenheim, 1968, 1969, 1973).

Although mono- and sesquiterpenoids are present in legumes, they have not been studied as intensively as they have been in confierous resins. On the other hand, considerable work has been done on elucidating the structure of the diterpenoids. Most of these analyses have been approached from the viewpoint of the structural chemist's interest in novel compounds, although interest in biogenetic interpretation relevant to systematics has been indicated (Ponsinet, Ourisson, and Oehlschlager, 1968). Harborne (1971) pointed out that about forty diterpenoids which have been isolated and characterized show considerable promise as taxonomic markers and that intensive systematic surveys of these compounds should be "especially rewarding." Diterpene resin acids, which Sandermann (1962) thought to have chemotaxonomic potential, have received the most study, although data are generally too scattered in terms of specific compounds taken from different organs and from isolated species to be of systematic and phylogenetic value as yet.

RESIN SECRETORY SYSTEMS AND CHEMISTRY IN *HYMENAEA*

Resin in *Hymenaea* is secreted by two anatomical mechanisms and resins from different organs are characterized by different suites of sesquiterpenoids and diterpenoids (Langenheim, 1967, 1969, 1973). Resin composed predominantly of sesquiterpenes is synthesized in parenchyma cells. These cells form an epithelial layer surrounding pockets, which are formed by schizogeny, in leaves, stipules, petals, sepals, and cortical and medullary tissue in young stems. In the trunk or root of the mature tree, resin consisting primarily of diterpenoids and some sesquiterpenes is secreted first into pockets which then form cavities produced by breakdown (lysigeny) of the resin secretory cells in the cambial zone. As a result of this lysigeny, large quantities of resin may accumulate and are exuded to the outside when a natural or man-made opening occurs in the bark. In pods the resin may be secreted initially in schizogenous pockets, but these may apparently be increased by lysigeny when the fruit is injured. Diterpenoids, again with some sesquiterpenes, constitute the resin of the fruits. Studies of resins of the leaves, trunk, and pods of *Hymenaea* are being carried out as a part of the coordinated study of the evolution of *Hymenaea* as a resin-producing genus.

Leaf Pockets: Leaf pocket sesquiterpenes of eleven species of *Hymenaea* (including thirty-five populations) have been analyzed quantitatively on a population or specific basis (Langenheim, Lee, and Martin, 1973; Martin, 1973; Martin, Langenheim, and Zavarin, 1974). Gas chromatographic analysis shows that this resin is composed primarily of eleven sesquiterpenes: caryophyllene and the α and β isomers of selinene are the major constituents in almost all *Hymenaea* species investigated, with lesser amounts of α- and β-copaene, α-cubebene, β-humulene, γ- and δ-muurolene, and a few minor unidentified compounds (Martin, Langenheim, and Zavarin, 1972, 1973, 1974).

Within an individual plant, quantitative composition of leaf pocket resins is closely maintained and is little influenced by position of the sampled leaf on the plant or season of the sampling (under greenhouse conditions). Intraplant stability of resin composition was further indicated by very low intraclonal variation in clones from two populations of *H. courbaril* grown under extremes of environmental conditions (Stubblebine, 1972, unpubl. data). In general, samples from eight greenhouse-grown plants per population were analyzed by Martin, Langenheim, and Zavarin (1973, 1974). Although in most studies of quantitative variation of resin composition a greater number of samples would be desired, in most of the populations of *Hymenaea* compositional stability is such that these numbers were statistically adequate.

Among *Hymenaea* species or between populations of certain species, quantitative leaf pocket resin compositional differences exist, superimposed upon the basic qualitative similarity of all members of the genus. Despite large quantitative compositional differences occurring within the range of certain species, interspecific differences in some instances permit chemosystematic interpretations. Investigation of the quantitative compositional data by numerical methods (Sokal and Sneath, 1963), including calculation of a variance-weighted coefficient of similarity (Flake and Turner, 1968) between every populational pair and clustering of these coefficients by the single linkage method, revealed groups of chemically allied populations and species (Martin, 1973; Martin, Langenheim, and Zavarin, 1974). For example, such analysis yielded a pattern of several species diverging from a large group of chemically similar species. The most divergent from this nuclear group are the African *H. verrucosa* and Amazonian *H. oblongifolia*, considered to be the most primitive species and derived from a common African ancestral stock in the early Tertiary. Also clearly divergent are the Amazonian *H. reticulata* (with its very restricted distribution there) and *H. velutina* from the *caatinga* of northeastern Brazil. *Hymenaea courbaril* is a member of the nucleus of closely related species, a relationship that fits many of our ideas regarding possible adaptive radiation within the genus.

In certain species or populations, however, chemosystematic value of leaf pocket composition is complicated by an unusually high degree of quantitative variability. In these groups, three compositional types could be distinguished based upon relative proportions of α- and β-selinene, caryophyllene and, the sum of γ-muurolene and δ-cadinene. Differences between these compositional types in a given population were sufficiently great that their overall quantitative compositions could be considered bimodal (trimodal in one case). For example, *H. parvifolia, H. aurea,* three populations of *H. courbaril* var. *courbaril,* and three populations of *H. courbaril* var. *stilbocarpa* show bimodal leaf-pocket resin composition. In all of these cases where bi- or trimodality has been found, either unusual morphological variability has been noted or the populations appear to be isolated as relicts. It is also interesting that all populations displaying this compositional diversity are limited to South America. Meso-American populations of *H. courbaril* var. *courbaril,* as well as other South American species, show relatively low variability in leaf pocket resin composition.

Trunk resins: Diterpenoids, particularly resin acids, are of considerable importance in *Hymenaea* trunk resins, comprising 45 to 85% of the soluble, nonpolymeric material. As mentioned previously, about forty diterpenoids have been isolated and characterized within members of the Leguminosae but only this study of *Hymenaea* has attempted to study them within organs from all species within a genus. In no previous study has there been a biological context nor an effort to put the analysis into a systematic or evolutionary framework. The structure of eight trunk resin acids has been elucidated for *Hymenaea verrucosa* (Hugel, Oehlschlager, and Ourisson, 1966; Martin and Langenheim, 1974). Two others have been isolated and characterized from *H. oblongifolia,* one from *H. parvifolia* (Cunningham, Martin, and Langenheim, 1973), and two additional ones from *H. courbaril* (Nakano and Djerassi, 1961; Cunningham, Martin and Langenheim, 1974), all from the Amazonian region. All resin acids but one for which we have structural information to date are bicyclic, belonging to the enantio-labdane skeletal series. Interestingly, *H. courbaril* has yielded both copalic acid and the antipodal labdane-13-en-8-ol-15-ioc acid. Martin and Langenheim (1974) compared the resin acids in ten species of *Hymenaea*, using gas chromatography and mass spectrometry, and found large quantitative differences among species. However, a number of the structures remain to be characterized and thus the total resin composition still cannot be fully assessed. Additionally, all work thus far has been done on resin samples which have been exuded to the exterior of the tree and thus exposed to the atmosphere for varying (and generally unknown) periods of time. It is not understood what compositional alterations and interconversions may occur during this period; consequently, quantitative comparisons of exuded trunk resin composition are not warranted. Presently studies of the oxidation-polymerization kinetics of the resin as it exudes from the tree are being made and may aid in later interpretation of the diterpenoid resin-acid data from the trunks of *Hymenaea.*

Pod resin: Diterpenoid resin acids are the primary constituents of pod resins in *Hymenaea.* Extensive analyses have been made of the resin acids from seed pods of *H. verrucosa,* showing them to be principally tetra- and pentacyclic diterpenes related to (-)kaurane (Hugel and Ourisson, 1965; Hugel *et al.,* 1965a, 1965b). Recently, Khoo, Oehlschlager, and Ourisson (1973) have isolated and determined the structure of four bicyclic pod-resin acids in *H. courbaril* from Guadelupe. However, these bicyclic resin acids are more highly rearranged than those isolated from the trunk resins.

The difference in structure from trunk to pod resins is of interest, but perhaps even more so is the skeletal difference between the pods of *H. courbaril* (rearranged bicyclic labdane) in the New World and *H. verrucosa* (tetracyclic kaurane and pentacyclic trachylobane) in the Old World. Differences of this degree had not appeared in either the leaf sesquiterpenes or the trunk diterpene resin acids.

GEOGRAPHICAL DISTRIBUTION AND ECOLOGICAL CONSIDERATIONS

The genus *Hymenaea* has an amphi-Atlantic distribution (Fig. 7). In the Old World *Hymenaea* is represented by the single species, *H. verrucosa.* It is restricted to the east coast of tropical Africa and adjacent islands, occurring in dry lowland evergreen forest, semideciduous woodland, and coastal evergreen bushland (Brenan, 1967).

In the New World thirteen species are distributed from central Mexico, through Central America, the West Indies, and all South American countries except Uruguay and Chile (Langenheim, 1973). They occupy all major tropical ecosystems along a moisture gradient from the equatorial rain forests in the Amazon and adjacent basins to the semidesert thorn forests of Mexico and northeastern Brazil (Fig. 8, Table 3). The evergreen forest in the Amazon Basin (*hylaea*) has several distinctive habitats with local names: (1) *igapó,* a permanently inundated swamp area, (2) *várzea,* a periodically inundated area, and (3) *terra firme,* an upland dry area. Evergreen forests also occur along the Atlantic coastal areas of Brazil from Rio Grande do Norte to the region around Rio de Janeiro, and the western coasts of Ecuador and Colombia north through Central America. Where a dry season of sufficient length occurs, evergreen forest is replaced by semideciduous and deciduous forests which, with increasing aridity, grade into savannahs or the arid thorn forests. In the savannahs of central Brazil (known locally as *cerrado*), evergreen forest trees from the Ama-

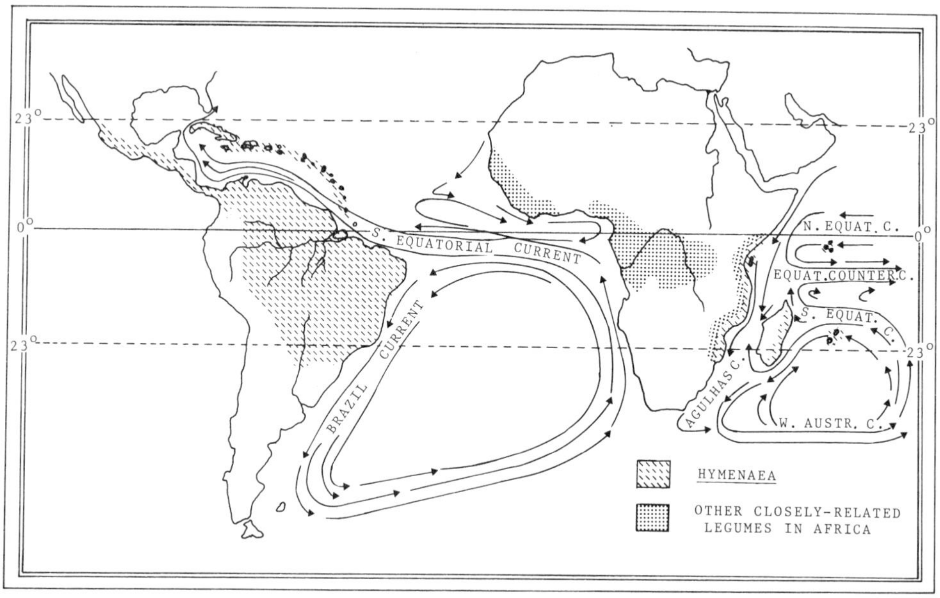

7. General distribution pattern of *Hymenaea* and other closely related legumes in Africa. Equatorial current patterns also indicated.

TABLE 3.
Distribution of Species of *Hymenaea* in Different Ecosystem Types with Sections and Groupings by Seedling Types Indicated.

Section	Seedling type	*Hymenaea* species	Ecosystem Type: Evergreen Forest: Amazonian: Terra Firme	Evergreen Forest: Amazonian: Várzea	Evergreen Forest: Amazonian: Ipagó	Evergreen Forest: Coastal	Evergreen Forest: Gallery/Valley	Semi- & Deciduous Forest	Savannah	Thorn Forest	Coastal Scrub
Sect. *Trachylobium*	Seedling Subtype I (Short Epicotyl)	*H. verrucosa*				X		X			X
		H. oblongifolia									
		var. *oblongifolia*	X	X							
		var. *latifolia*				X					
		var. *palustris*			X						
		var. *davisii**	X								
		H. parvifolia	X								X
Sect. *Hymenaea*		*H. torrei**				X					
		H. intermedia									
		var. *intermedia*	X								
		var. *adenotricha**	X								
		H. reticulata	X								
		H. rubriflora									
		var. *rubriflora*				X					
		var. *glabra**				X					
		H. aurea			X						
		H. eriogyne								X	
		*H. maranhensis**							X		
	Seedling Subtype II (Long Epicotyl)	*H. courbaril*									
		var. *courbaril*	X	X	X	X	X	X	X	X	X
		var. *altissima*				X					
		var. *longifolia*							X		
		var. *villosa*				X					
		var. *stilbocarpa*					X	X			
		var. *subsessilis*	X	X							
		H. martiana					X		X	X	
		H. velutina								X	
		H. stigonocarpa									
		var. *stigonocarpa*							X		
		var. *pubescens*							X	X	
		var. *brevipetiolata**							X		

*Seedling type known by inference.

zon Basin extend along the river valleys (often called gallery forests). Because the largest area of thorn forest occurs in northeastern Brazil, we shall refer to it by the Brazilian name *caatinga.*

The density of individuals within species, as well as species, varies greatly with ecosystem type. Since density is important regarding breeding structure and thus understanding possible patterns in speciation, the few data available are presented here in reference to geographic distribution and relationship to ecosystem type.

Quantified sociological descriptions of tropical ecosystems are rare, and most of our information about population density of *Hymenaea* is based upon observations. Generally speaking the greatest degree of dispersion of the individuals within a species of *Hymenaea* occurs in rain-forest ecosystems. In a detailed sociological study in the Amazonian *hylaea,* Black, Dobzhansky, and Pavan (1950) and Pires, Dobzhansky, and Black (1953) found two individuals of an unidentified species of *Hymenaea* in a 3.5 hectare plot in one Amazonian *terra firme* site near Belém, Pará, but none in a relatively nearby one hectare plot. Two individuals of *H. parvifolia* and one individual of an unknown species of *Hymenaea* occurred in one hectare *igapó* plot along the Rio Guamá near Belém. These studies point out the high degree of dispersion of *Hymenaea* species and individuals within species in the communities occurring in these *terra firme* and *igapó* habitats in the eastern floristic zone of the Amazon Basin. These studies also indicate differences in composition between the two Pará *terra firme* plots less than 120 km. distant from each other. Of the eighty-seven species found in one plot and one-hundred seventy-nine in the other, only fifty-four were common to both. Because of the species recorded in one, but not in the other locality are known to be widely distributed in *terra firme* forests ofPará, the authors assume their absence to be "accidental." Only a few species occurred in both the *terra firme* and *igapó* forests where the plots were only 2 km. apart; however, the environmental conditions are strikingly different between *igapó* and *terra firme* habitats and hence selective factors are likely to be considerably different in them. In one locality, the Ducke Reserve near Manaus in the central Amazon, three species of *Hymenaea* appear in close proximity in a *terra firme* site: *H. parvifolia, H. intermedia,* and *H. reticulata.* This is the closest situation to what might be considered sympatric distribution of *Hymenaea* species that we have observed. *Hymenaea courbaril* var. *subsessilis* also occurs a short distance away, but tends to be restricted to sandy habitats.

The density is also low for the species which occur in the Atlantic coastal humid forests. *Hymenaea aurea* and *H. oblongifolia* var. *latifolia* occur infrequently in the "*Hylaea bahiana*" (Andrade-Lima, 1953), as does *H. rubriflora* farther north in Pernambuco and *H. courbaril* var. *altissima* south in Guanabara and São Paulo. An undescribed *Hymenaea* species again is highly dispersed in the rain forest of the Osa Peninsula of Costa Rica.

In semideciduous forests of Mexico (*Selva Alta* or *Mediana Subcaducifolia sensu* Miranda and Hernandez-X, 1963), *H. courbaril* is a characteristic tree but is not dense, except along rivers in southern Mexico where it forms *"selvas de guapinol"* (forests of *H. courbaril*). In certain stands of *Selva Mediana Subcaducifolia, e.g.,* along the coast in Guererro, Mexico, it becomes a dominant species, actually forming a consociation in some localities. It also here appears on sandy plains or beach ridges intermingling with *Selva Baja Caducifolia* (low deciduous forest).

Hymenaea courbaril also occurs in moist to relatively dry forest types in Costa Rica

(Tropical Dry Forest *sensu* Holdridge, 1964; Holdridge, *et al.* 1971), as well as being a common component of the Subtropical or Premontane Moist forests (*sensu* Holdridge, 1964). In both of these forests, the greatest density of individuals occurs along the rivers.

In the *cerrado* of central Brazil, *H. courbaril* var. *courbaril* and var. *stilbocarpa* also occur along the rivers. Increasingly southward, however, var. *stilbocarpa* replaces var. *courbaril.* Some authors, *e.g.*, Ratter, *et al.* (1973) make a distinction between gallery and valley forests, with the former being in more-or-less swampy sites (where the water table is close to the surface even during the dry season) and the latter being at a slightly higher level of the stream valleys. In a study of the northeastern part of the Mato Grosso near the contact of the *cerrado* vegetation characteristic of central Brazil and the Amazonian forest, Ratter, *et al.* (1973) found *H. courbaril* var. *stilbocarpa* restricted to the valley forest, with *H. stigonocarpa* var. *pubescens* frequent on flat, rocky ridges in this part of the *cerrado.* Our observations would support a similar situation for the *cerrado* in Goías. *Hymenaea martiana* also appears in local aggregations in some valleys in the *cerrado.*

In the thorn forest (*caatinga*) of northeastern Brazil, *H. velutina, H. eriogyne,* and *H. martiana* occur only sporadically throughout the *caatinga,* but they have a clumped distribution where they do occur. Likewise in our experience *H. velutina* and *H. eriogyne* are sympatrically distributed (Fig. 8f). *Hymenaea stigonocarpa* tends to be more evenly distributed and less frequent in the *caatinga* than in the *cerrado.*

PHENOLOGICAL OBSERVATIONS

Flowering and Fruiting Times

Information on time of flowering and fruiting for *Hymenaea* species comes from dates given on herbarium sheets, general field observations while collecting, and from a single, long-term, phenological study made at the Ducke Reserve near Manaus from 1962 to 1972 (Araujo, 1970; Volpato and Schmidt, pers. comm.). Although flowering is known to occur throughout the year in many tropical areas, particularly in equatorial rain-forest ecosystems, some workers have attempted to correlate flowering time with seasonal rainfall distribution. For example, Ducke and Black (1953) indicated from their extensive collecting and observations in the Amazon Basin that flowering of trees in the *terra firme* forests, where there is a long, heavy rainy season, is restricted to the beginning or end of the dry season. In those parts of the rain forest where the two seasons of the year are not sharply defined and are often irregular, flowering occurs every month. In all of the ecosystems in which various species of *Hymenaea* occur, the flowering time generally falls within the dry season or at the beginning of the rainy season.

Fruit of *Hymenaea* appears to attain full size two months after pollination and fertilization are completed. The fruit stays on the tree for at least six to eight months before it falls. Although the fruit generally reaches full size early in its development, apparently maturation of seeds (including enlargement of cotyledons and hardening of testa) only takes place two or three months before the fruit falls.

In a study of seed predators of *H. courbaril* in seasonally dry forests, Janzen (1969, 1971b) pointed out that trees in dense forest stands appear to have a different reproductive pattern from those in more open sites. Those in the dense habitats have a several-year flowering and fruiting period, whereas those in the open habitats may reproduce in consecutive years. Janzen suggests that the adaptive significance of sexual dormancy is in strongly

8. Photographs of *Hymenaea* species in various habitats in Brazil. a. *H. oblongifolia* var. *palustris* in *igapó* along river bank, Serra do Nevio, Amapá. b. *H. oblongifolia* var. *latifolia* along river in coastal evergreen forest near village of Rio Branco, Bahia. c. *H. reticulata* in *terra firme* rain forest, Ducke Reserve, Manaus, Amazonas. d. *H. courbaril* var. *courbaril* in *terra firme* rain forest, Serra do Nevio, Amapá. e. *H. courbaril* var. *longifolia* in ecotone between *caatinga* and *cerrado,* border of Chapada do Araripe, Ceará. f. *H. eriogyne* in *caatinga* near Picos, Piauí (note its bare branches, *H. velutina* in background in full leaf). g. *H. martiana,* on the edge of *caatinga* SW border of Ceará and Piauí. h. *H. stigonocarpa* var. *pubescens* in *cerrado* near Brazilia, D.F.

reduced predation by *Rhinochenus* beetles. A two-year cycle of flowering for *H. intermedia* has also been noted in the Ducke Reserve in the central Amazonian region (Araujo, 1970; Volpato and Schmidt, pers. comm.).

The evidence available from most of the species restricted to the Amazon Basin (*H. oblongifolia* vars. *oblongifolia* and *palustris, H. intermedia* vars. *intermedia* and *adenotricha, H. reticulata,* and *H. parvifolia*) indicates flowering from August to November although in some species, flowering either begins a month earlier or lasts a month later (Table 4). The only exception to this pattern is the June flowering of *H. oblongifolia* var. *davisii,* known only from the type collection from Guyana. Thus, the flowering generally occurs during the dry season or in the transition to the wet season (Langenheim, Lee, and Martin, 1973).

In the Atlantic coastal forests, *H. aurea* and *H. oblongifolia* var. *latifolia* flower in a relatively dry period from December through February, before the heavy wet period from March to July (Table 4). They do not, however, flower in the driest period from August to September. The flowering period of *H. rubriflora* in Pernambuco starts toward the end of the dry season in December but continues well into the wet season in May (Table 4).

TABLE 4.
Time of Flowering of Various *Hymenaea* Species and Varieties
(*H. courbaril* excluded)

Hymenaea species	General localities	Months											
		J	F	M	A	M	J	J	A	S	O	N	D
H. verrucosa	E. coastal Africa & adj. islands		*X*	*X*	*X*								
H. oblongifolia													
var. *oblongifolia*	Amazon and adjacent basins								*X*	*X*	*X*		
var. *davisii*	Guyana						*X**						
var. *latifolia*	Southern coastal Brazil	*X*	*X*										*X?*
var. *palustris*	Amazon and adjacent basins									*X*	*X*	*X*	
H. parvifolia	Amazon and adjacent basins									*X*	*X*	*X*	*X*
H. intermedia													
var. *intermedia*	E. & Central Amazon Basin								*X*	*X*	*X*		
var. *adenotricha*	W. Amazon Basin							*X**	*X**				
H. reticulata	Central Amazon Basin									*X*	*X*	*X*	
H. torrei	Cuba						*X**						
H. rubriflora													
var. *rubriflora*	Coastal Brazil	*X*	*X*	*X*	*X*	*X*		*X**					*X*
var. *glabra*	Coastal Brazil		*X**										
H. aurea	Coastal Brazil	*X*	*X*										*X*
H. stigonocarpa													
var. *stigonocarpa*	Central & S. Brazil & Paraguay	*X*	*X*									*X*	*X*
var. *pubescens*	Central & S. Brazil & Paraguay	*X*	*X*	*X*									*X*
var. *brevipetiolata*	Southern Brazil & Paraguay												*X**
H. maranhensis	Maranhão, Brazil		*X**		*X**								
H. martiana	Central and south Brazil									*X*	*X*	*X*	*X*
H. velutina	Northeastern Brazil										*X*	*X*	*X*
H. eriogyne	Northeastern Brazil		*X*	*X*									

*Information derived from a single specimen.

However, one specimen of *H. rubriflora* from the Rio Grande do Norte recorded flowering in July.

The other rain-forest species, *H. verrucosa* from coastal east Africa, flowers from February to April at the end of the dry season and beginning of the wet one. The Cuban *H. torrei,* according to the data on the type collection, flowers in June.

In the *caatinga* areas of northeastern Brazil, *H. velutina* flowers from October to December, again in the dry season. *H. eriogyne,* which grows sympatrically with *H. velutina* in several localities in southern Piauí, appears to flower about two months later, just before the onset of the short wet season in this semidesert region (Table 4). *Hymenaea maranhensis,* which occurs in the northeastern Brazil *cerrado,* also flowers on this later schedule (February to April).

Hymenaea martiana flowers from September to December whereas *H. stigonocarpa* and its varieties begin flowering around November and continue through March. Although the flowering period of these *cerrado* species of central and southern Brazil falls about two months later than that of the species of the Amazon Basin, the correlation is still essentially with the dry season.

Because the distribution of the polymorphic *H. courbaril* is essentially that of the genus in the New World, flowering times are different among various populations along the geographic gradient. In all cases the flowering time is either coincident with the dry season or occurs during the transition to the wet season. *Hymenaea courbaril* flowers from March to May in Mexico, the West Indies, Central America, and the northern countries of South America (Table 5). In the Amazonian rain forest, var. *courbaril* flowers from August to November, whereas var. *subsessilis* flowers from October to December. The differential flowering time of the two varieties cannot yet be explained fully in terms of climatic pattern. However, when a number of species or, in this case, varieties of flowering plants with sequential flowering times occur together, it might make possible the sustained nutrition of large pollinators in a geographical region (Allen, 1939; Van der Pijl, 1956a; Baker, 1963, 1970, 1973). Such a flowering complex should be particularly advantageous when the number of species of potential pollinators is low and when pollinators are long-lived (Heinrich and Raven, 1972). *Hymenaea courbaril* var. *courbaril* in northeastern Brazil flowers from October to November, a pattern similar to that of varieties *villosa* and *longifolia* growing in the same general area. The two southern varieties, *altissima* and *stilbocarpa,* flower later than the other varieties of *H. courbaril* just discussed: var. *stilbocarpa* from October to December, var. *altissima* from December until late February (Table 5).

LEAF-DROP

Although *Hymenaea* is usually considered to be evergreen, most species seem to be facultatively deciduous and one or two seem to be obligately deciduous. Trees may go through a heavy leaf-fall during dry periods and occasionally lose all their leaves under severe moisture stress (Langenheim, Lee, and Martin, 1973). Again, Araujo's (1970) phenological data for *H. intermedia* var. *intermedia* (with general observations also for *H. parvifolia* and *H. reticulata*) from Ducke Reserve give important leads to behavioral activity. Apparently leaf-fall varies with the flowering-fruiting cycle. In years of flowering only, leaf-fall immediately precedes flowering; in years of fruiting, leaf-fall occurs when flowering would normally take place. In both cases, however, the leaves fall during the dry period from June through September. Although many leaves fall and new ones are put

TABLE 5.
Time of Flowering of *H. courbaril*

Hymenaea courbaril	General localities	Months											
		J	F	M	A	M	J	J	A	S	O	N	D
	Mexico			X	X	X							
	West Indies			X	X	X	X						
	Guatemala				X	X	X						
	Honduras			X	X	X							
	Costa Rica			X	X								
	Panama			X	X	X							
var. *courbaril*	Venezuela			X	X	X							
	Colombia				X*							X*	
	Guyana			X	X								
	Surinam					X	X						
	Amazonian Brazil								X	X	X		
	Northeastern Brazil										X	X	
var. *subsessilis*	Amazonian Brazil										X	X	X
var. *longifolia*	Northeastern Brazil											X	X
var. *villosa*	Coastal Brazil											X	X
var. *altissima*	Central & S. Brazil	X	X										X
var. *stilbocarpa*	Central & S. Brazil										X	X	X

*Information derived from a single specimen.

out each year, trees are rarely completely denuded. The possible exceptions are *H. courbaril* var. *subsessilis,* essentially restricted to sandy beaches along the central Amazon, which apparently does become completely defoliated from August to September, and *H. eriogyne,* found only in the *caatinga* of northeastern Brazil, which loses all its leaves for a considerably longer period of time. Several populations of *H. eriogyne* were completely denuded in northern Piauí in November.

We have not observed a pronounced leaf-drop in *H. oblongifolia, H. courbaril* (except var. *subsessilis* and occasionally var. *courbaril*), or in *H. parvifolia.* In the Atlantic coastal forest, *H. rubriflora* and *H. aurea* clearly lose many leaves during the dry season. During these periods *H. oblongifolia* var. *latifolia* still possesses essentially its full complement. In the interior, *H. martiana* and *H. stigonocarpa* lose leaves continuously, but remain evergreen.

Hymenaea verrucosa is reported to be an evergreen tree. No information is available for either *H. torrei* or *H. maranhensis.*

In the seasonally dry forest of Central America, Janzen (1967) noted the general correlation of leaf shedding with a period preceding the peak of flowering and also observed maximum activity of pollinating insects during the dry season, leading to an efficient pollination system. He hypothesized that the relationship between loss of leaves and peak of flowering probably resulted from the presence of insects or other pollinating agents during the dry season (at the most opportune time of the year) rather than from physiological processes which could only occur at that time of the year. Whatever leaf-drop occurs usually precedes flowering in *Hymenaea,* but the relationship of this sequence to pollinating vectors in the different ecosystems is not yet understood.

REPRODUCTIVE BIOLOGY

The nature of a breeding system is central to an understanding of speciation. Unfortunately, it is impossible to experiment with this system in a tree such as *Hymenaea* in a short period of time, and we thus are dependent upon analysis of relatively few observations and inferences from them.

There has been considerable discussion of the fundamental problems as to whether random drift or natural selection dominates in populations of tropical trees (Dobzhansky, 1950; Corner, 1954; Stebbins, 1957, 1958; Baker, 1959; Snow, 1966; Federov, 1966; Ashton, 1969; Bawa, 1974). Corner (1954) first suggested that rain forest trees might be self-pollinated because they flower asynchronously and he thought that the movement of their pollinators must be limited to small distances. Baker (1959) also pointed out that inter-tree movement of pollinators might be inadequate in the structurally complex rain forest, thus leading to self-pollination and inbreeding. Federov (1966) then strongly advocated the view that the effective breeding population is so low in tropical rain-forest species (due to spatial isolation resulting from low densities of individuals) that self-fertilization "must be the rule." This situation would then result in the dominance of "automatic genetic processes," such as genetic drift, over processes of natural selection, leading to rapid evolution through increase in homozygosity and nonadaptive differentiation of small populations. It is also argued by the proponents of this idea that the physical environment of the tropical rain forest is so unvarying that there are not likely to be any operative differential selective forces. Thus, species have not adapted into specific niches.

Ashton (1969), however, believes that, among various genera in the Dipterocarpaceae in tropical southeast Asia, outcrossing is probably predominant between individuals of a clump as well as between clumps. He further agrees with Dobzhansky (1950) that in the rain forest there are highly specialized adaptations to the physical environment (particularly edaphic ones, since the temperature and humidity generally remain constant in equatorial habitats) and to the extremely complex biotic environment. In other words, adaptations will be directed toward greater integration with edaphic and biotic factors of the environment. Nonetheless, Poore (1968) has also shown in the Dipterocarpaceae that an apparent single microhabitat may be filled with one or several alternative species by chance. At the time of formation of a "gap" in the forest, saplings of a particular species by chance might be at just the right stage to take their place in the canopy.

Janzen (1970) has presented the view that habitats, such as those in tropical rainforests, will support more species of trees as they become more favorable for seed and seedling predators. Predator pressures prevent any one species from becoming sufficient ly abundant to outcompete the others. Janzen points out that the challenge of predators can be met only through behavioral or chemical changes, which must be transmitted genetically. The more favorable the physical environment to the predators, the more frequently in evolutionary time the chemical defenses of the plant will have to be modified through genetic change if the plant is to be successful in the community. Janzen further suggests that these circumstances dictate the importance of outcrossing. Bawa (1974), however, indicates that although it appears that tropical forest trees are largely cross-fertilized, inbreeding is still possible. Since rain forest species have low population densities, inbreeding to a certain extent is expected if gene flow is limited to small distances. In an investigation of the breeding system of one hundred thirty species in a tropical lowland semideciduous forest in Costa Rica, Bawa (1974) found using controlled pollinations, only 14% of the species were self-compatible, whereas 54% were self-incompatible, 22% dioecious, and 10% monoecious. Unfortunately, a similar study has yet to be done in the tropical rain forest.

Bawa and Opler (1975) also have pointed out that in tropical lowland forest one-fourth to one-half of all species have unisexual flowers and a majority of these species are dioecious. They postulate that in the majority of species investigated dioecism has probably evolved from a self-compatible breeding system in response to selective pressure for outcrossing.

Although these discussions are generally restricted to investigations of tropical lowland forests (both rain forest and semideciduous forests), similar sorts of considerations exist in the savannah and thorn forests.

In the absence of experimentation, our analysis of *Hymenaea* has to rest upon a series of inferences. Although data are scattered, *Hymenaea* does appear to provide support for the role of selection and adaptive radiation from rain forest ecosystems with a probably outcrossing system. The effective size of outcrossing populations may be influenced by density, continuity of distribution, and reproductive periodicity, all of which have been discussed in previous sections. Now we shall discuss the floral adaptations which appear to favor cross-pollination and the availability of pollinators which might work over relatively long distances. The relationships of possible hybridization inferred from morphological variation, polyploidy observed in cytological studies, and seed dispersal to speciation are also considered.

TABLE 6.
Inflorescence and Floral Differences of Sections *Trachylobium* and *Hymenaea*

	Sect. *Trachylobium*	Sect. *Hymenaea*
Inflorescence Type	Long-paniculate	Densely corymbose
Nature of Branches	Long, slender, & slightly flexuous	Short, stout, & strongly flexuous
Number of Flowers per Inflorescence	Many (to 200 +)	Fewer (to 70 +)
Flower Size	Small (older buds ca. 14 x 7 mm.)	Medium to large (older buds 18 x 9-35 x 18 mm.)
Flower Color	Creamy white or rose-tinged	Creamy white (rarely red)
Presence of Flower Odor	+?	+
Presence of Disc	+	+
Presence of Additional Nectariferous Tissue	–	– or +

Inflorescence, Floral Structure, and Behavior Relevant to Pollination Mechanisms

There are major differences in the inflorescence and floral structure of the two sections of *Hymenaea* (Table 6) which affect the possible type of pollinator. The flowers of sect. *Trachylobium* are rather small, numerous, and arranged at the periphery of a conical panicle, whereas in sect. *Hymenaea* the flowers are much larger, fewer, and reach about the same level during anthesis. In the latter case, their axes are pendant to horizontal and their ramifications almost always face outward, so that the inflorescences are more or less one-sided. Another significant difference between the sections is the presence of additional nectariferous tissue within the hypanthium in the members of sect. *Hymenaea* which is absent in sect. *Trachylobium.*

Thus far, no information has been obtained about the floral behavior for species of sect. *Trachylobium.* The following discussion therefore is limited to species in sect. *Hymenaea,* as primarily exemplified by *H. courbaril.*

The flowering period of an inflorescence of *H. courbaril* is weeks long, even though the flowers are short-lived. Vogel (1968-1969) noted that only two to five flowers of the approximately seventy in an inflorescence of *H. courbaril* var. *subsessilis,* from near Manaus in the central Amazon, are open at any one time. Similarly, Frankie (pers. comm.) observed in *H. courbaril* var. *courbaril* in Guanacaste, Costa Rica, that, although the trees produced large flushes of flowers, they produced some flowers nightly for long periods. Both Vogel and Frankie indicated that the flowers of *H. courbaril* open near sunset, usually between 5 and 9 P.M. and lose their calyx, corolla, and stamens by the following noon. Further observations on the floral behavior have been made for *H. courbaril* in Costa Rica by Frankie (pers. comm.) and by YTL for *H. courbaril* var. *subsessilis* and *H. reticulata* near Manaus, Amazonas, Brazil.

In all of these cases the calyx lobes are strongly imbricate in bud and are borne on the margin of the turbinate-campanulate hypanthium. The calyx lobes and five large, creamy white, slightly over-topping petals open only halfway during anthesis (Vogel, 1968-1969). The ten exserted stamens, with relatively large, pollen-rich anthers, are inserted on an elevated disc around the inner rim of the hypanthium. The stipe of the ovary is attached laterally to the inner wall of the hypanthium supporting the ovary, and the involute style above.

The anthers are relatively large compared to the size of the flower. In bud, the dorsifixed anthers and the folded filaments are in two whorls around the central pistil, and occupy most of the space inside the flower (Fig. 9 a,b). The anthers are fleshy and the supply of pollen is copious. Frankie observed bats boring into mature flowering buds of *H. courbaril* with their tongues, probably to obtain pollen for food.

At the beginning of anthesis, the filaments of the ten stamens start to unfold and exsert the anthers (Fig. 9 c-e). It probably is this thrusting force in the unfolding process that results in the opening of the calyx and corolla. At this stage the intact anthers project beyond the perianth but the stigma remains protectively coiled inside the corolla (Fig. 9 f). The projecting anthers become fully extended, shed their contents, and dehisce within a couple of hours of the opening of the flower. Later, the style starts to uncoil. In an unopened flower, the small capitate stigma always faces upward and overtops the stamens (Fig. 9g). Frankie (pers. comm.) thinks that the stigma might be protected from self-fertilization by remaining coiled inside the corolla, assuring outcrossing if the pro-

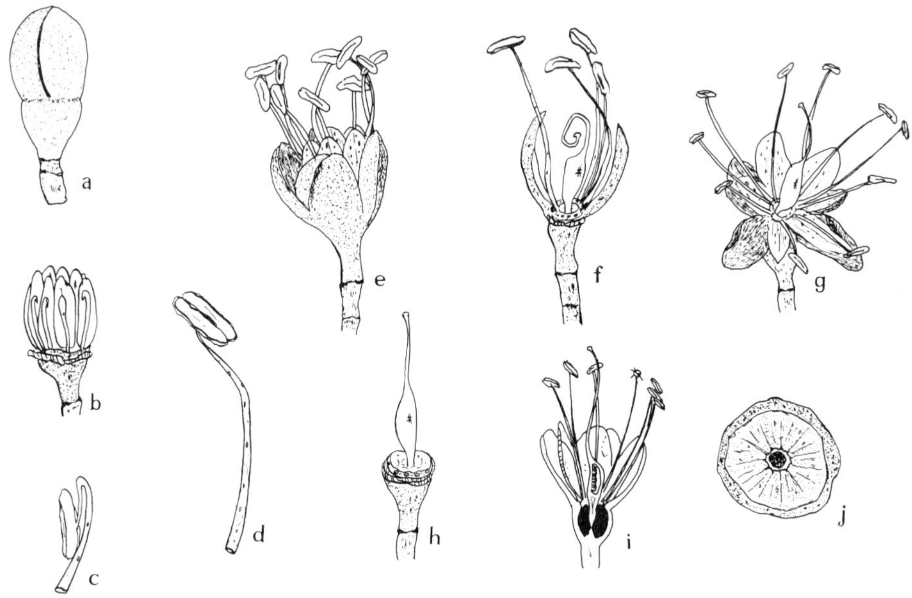

9. Flower of *H. courbaril*. a. flowering bud shortly before anthesis. b. flowering bud with sepals and petals removed showing stamens. c. stamen in bud showing folded filament. d. stamen. e. partly open flower. f. partly dissected flower showing involute pistil. g. fully open flower. h. partly dissected flower showing pistil and disc. i. longitudinal section, showing additional nectariferous tissue within the hypanthium (darkened portion). j. cross-section of hypanthium showing nectariferous tissue. (e-g, drawn from photographic slides courtesy Dr. Frankie; i, redrawn from Vogel, 1968-9.)

tected stigma were receptive at this time. Self-incompatibility has been reported for *H. courbaril* in the dry forests of Guanacaste, Costa Rica (Bawa, 1974).

During anthesis the hypanthium is filled with nectar. The quantity of nectar is relatively small compared to that secreted by flowers from such a plant as *Parkia* (Vogel, 1968-1969; Baker, 1973). In *H. courbaril* and other species of sect. *Hymenaea* the production of relatively few flowers daily over a period of weeks correlated with a quantity of nectar, which is attractive but not satiating, corresponds well with flower visitors flying from plant to plant, the "trapline syndrome." This type of behavior should greatly increase chances of outbreeding even in highly dispersed populations of *Hymenaea.*

Differences between types of inflorescence and floral structure (Table 6) strongly suggest that different major pollinating agents may be involved in the two sections. Unfortunately, information is not yet available on pollination mechanisms in sect. *Trachylobium,* as was the case with floral behavior. Observations only have been made on members of sect. *Hymenaea,* again in the species *H. courbaril* in Central and South America.

Hymenaea courbaril has characters that fit it well to the bat-pollination syndromes described by various workers (Van der Pijl, 1936; Baker, 1961, 1973; Vogel, 1968-1969); (1) nocturnal anthesis; (2) flowers with a strong odor; (3) short, stout, and strongly flexuous inflorescence branches which bring most open flowers to the same level; (4) presence of massive nectariferous tissue and retention of nectar in a broadly tubular corolla; and (5) a pattern of anthesis which promotes pollination by trap-lining vectors. Indeed, active bat visitation of *H. courbaril* has been reported by several workers (Table 7). Carvalho

TABLE 7.
Bat Visitation of *H. courbaril* in Costa Rica and Brazil

Plants	Flowering season	Observed bat visitation
Hymenaea courbaril var. *courbaril* Belém, Brazil	Aug.-Sept.	*Glossophaga soricina* (Carvalho, 1961)
Hymenaea courbaril var. *subsessilis* Manaus, Brazil	Oct.-Dec.	*Phyllostomus discolor* (Vogel, 1968-1969)
Hymenaea courbaril var. *courbaril* Cañas, Costa Rica	Mar.-Apr.	*Sturnira lilium* *Carollia "perspicillata"* *Artibeus jamaicensis* *Glossophaga soricina* *Artibeus phaeotis* *Phyllostomus discolor* (Hcithaus, *ct al.* 1974)

(1960, 1961) observed bat activity in the eastern Amazonian region. Vogel (1968-1969), in a detailed study and review of bat pollination in the New World, noted active visitation of bats soon after sunset to many flower clusters of *H. courbaril* var. *subsessilis* in the central Amazon. Although he observed that bees (Meliponidae) and probably humming birds (Trochilidae) vigorously exploited *Hymenaea* flowers before sunset and that sphinx moths and night moths also participated in pollen transmission after sunset, he considered that bats were the most effective. He described the timing, flight behavior, and contacts made by bats with the flowers. Heithaus *et al.*, (1974) also recorded six species of bats visiting flowers of *H. courbaril* in Costa Rica. It has also been confirmed that they transmit pollen.

Vogel (1968-1969) indicated that bats might visit and pollinate all thirteen Brazilian *Hymenaea* species listed by Ducke (1935) instead of just *H. courbaril.* We think, however, that bat pollination is likely only among those species that have inflorescences and flowers similar to those of *H. courbaril* (sect. *Hymenaea*). Species in sect. *Trachylobium,* on the basis of inflorescence and floral characters, may well be pollinated by insects. In the Guamá Reserve, IPEAN near Belém, Brazil, collections of insects visiting a tree of *H. oblongifolia* var. *oblongifolia* during early morning hours have been made from towers built near it. A large number of bees and wasps have been collected, but they have not yet been identified, nor have they been scrutinized for pollen (Langenheim, Lee, and Martin, 1973). On several occasions, YTL observed flowers of *H. oblongifolia* vars. *palustris* and *latifolia* being visited by bees and butterflies. Further, support for the possibility of predominance of insect pollination may be derived from *H. verrucosa,* the lone *Hymenaea* species of the Old World. It is assumed that since bats in the paleotropics all belong to suborder Megachiroptera, these relatively large and inactive bats could hardly be involved in pollination of *H. verrucosa* because of the fragile inforescences and relatively small flowers. Assuming that New World *Hymenaea* species have been derived from an African ancestral stock (Langenheim, 1973; Langenheim and Lee, 1974), chiropterophily among them may then be of secondary and relatively recent origin. Vogel argues that the less specialized morphology and "exposed" nectary of *Hymenaea* represents a more primitive type of chiropterophilous flower, which depends upon short-tongued Phyllostomids. This does not, of course, exclude visitations by Glossophaginae as indicated by Carvalho (1960, 1961) and Heithaus *et al.* (1974).

Seed Dispersal

Seed dispersal is another important mechanism for maintenance of genetic diversity in widely dispersed populations. As Grant (1956) pointed out, the dispersal of one seed is potentially equivalent to dispersal to a similar distance of thousands of pollen grains in terms of eventual genetic results. Many plants adapt to the deep shade of the rain-forest floor by evolving heavy fruits, enclosing sufficiently large seeds that ample food reserves are available for a young sapling to become established even with little light. Such fruit often present problems for effective dispersal.

Seeds of *Hymenaea* are generally dispersed within the woody, indehiscent fruits which remain intact. Water is one of the major ways in which the fruits, and enclosed seeds, are dispersed. Fruits are frequently seen floating intact in rivers, and this may partially account for the density of various species of *Hymenaea* along streams or rivers in all ecosystems (including the *igapó* and *várzea* in the Amazon Basin). *Hymenaea* pods are also oceanically dispersed. Pods from the West Indies commonly are found along the southern shores of Florida (Gunn, 1968). Some of these pods recently were furnished to us by Gunn, and the seeds germinated rapidly (from 7 to 10 days).

Seeds are also dispersed by the activities of rodents and peccaries. As previously indicated, the seeds are mature when the pods fall to the ground. The fruits fall during the rainy season and may thus be relatively quickly attacked by soil microbiota. The seeds occasionally germinate within the decaying fruit. In addition, various mammals crack open the pods, seeking particularly the sweet, pulpy endocarp, and generally leaving the seeds behind. They sometimes carry the pods off and may eat some of the seeds while consuming parts of the fruit. It is thought that the seeds pass through the alimentary tract and may even be scarified by stomach acids. With the aid of such mammals, seeds may be transported some distance from the parent tree.

Van der Pijl (1956b), in a survey of dispersal of leguminous seeds and fruits, indicated that the "juicy" fruits of *H. courbaril* are dispersed by tree mammals, and possibly also by ground mammals. We think that the woody, indehiscent fruits of *Hymenaea*, which contain floury endocarp, are not dispersed by tree mammals. Van der Pijl's conclusion is probably due to his erroneous translation of Burkart's (1952) general description of fruits of *Hymenaea*.

Hybridization, Polyploidy, and Apomixis

Interfertility of related species tends to be the rule among most woody angiosperms; however, evidence for hybridization is rare among rain-forest species. Ashton (1969) has indicated that there appears to be strong selective pressure against hybridization of trees in rain-forest ecosystems. He points out that few climax rain-forest species are able to colonize either naturally or artificially cleared sites successfully. It is here, in a new habitat and under reduced competition, that hybrids might have the greatest chance of survival. There also is little evidence of introgression.

Although there is no clear evidence for hybridization in *Hymenaea,* there still is to be considered the possible hybrid origin of *H. intermedia* from *H. courbaril* and *H. oblongifolia. Hymenaea intermedia* certainly possesses diagnostic characters of both these latter species; however, it seems more probable to us that *H. courbaril* evolved from *H. oblongifolia* stock via the intermediate *H. intermedia* populations. Also it appears possible from leaf pocket resin composition that there may have been gene flow (and possible introgressive hybridization) between an undescribed species of *Hymenaea* from the Osa Peninsula and *H. courbaril* in Costa Rica (Martin, Langenheim, and Zavarin, 1974).

There is no evidence of polyploidy in *Hymenaea.* In fact, all of the eleven species and nine varieties investigated recently show the chromosome number to be 2n = 24 (Table 2).

The uniformity of chromosome number in *Hymenaea* follows the general pattern of tropical leguminous trees in rarely exhibiting any variation in chromosome number. This stability is not merely restricted to species of the same genus, but is also commonly found at the tribal level (Atchison, 1951; also see Appendix A, Heywood, 1971), and may even extend to higher taxa, even to subfamilies. With members of Detarieae as an example, chromosome numbers had been prepared for fifteen genera (see Appendix A, Heywood, 1971). Among these genera, twelve (*ca.* 80%) have the same basic number x = 12; the remaining three genera (*ca.* 20%) have a different basic number x = 8, 10, or 11. Thus chromosome number is of little value for taxonomic discrimination at specific, generic, or tribal level in this instance. It also gives no indication that polyploidy has been directly involved in the evolution of these groups.

Smith-White (1959) has noted that evolution into open habitats among woody families is often associated with increase in variability in chromosome number. This obviously has not been shown by *Hymenaea,* although it has in *Guibourtia,* a closely related, resin-producing genus in Africa. For example, *G. chie,* a rain-forest tree, has the usual diploid number of 24, whereas the savannah species of *G. coleosperma* has 48.

Hymenaea can reproduce vegetatively; in Mexico and Costa Rica, JHL discovered that numerous saplings were not derived from seeds but were produced asexually from an extensively spreading root system. This development of root sprouts was particularly evident at the forest edge or where *H. courbaril* had been left isolated in fields cleared for agriculture. Also, in areas along the coastal dry, deciduous forests in Guerrero, Mexico, crown sprouts were common from stumps of trees.

INFERENCES REGARDING SPECIATION

Several generalities regarding the speciation process in *Hymenaea* may be pointed out. Considerable morphological constancy, in characters of flowers and fruits characterize most populations; there is little evidence of clinal variation even in the widely distributed populations such as *H. courbaril* and *H. oblongifolia.* This also is apparently true with the chemical characters of the resin. Morphological and chemical variability is manifest in a few species such as *H. parvifolia,* although there is no evidence for either hybridization or polyploidy.

The densities of individuals of particular species vary greatly in different ecosystems; however, pollinating agents seem adequate to maintain gene flow even among the highly dispersed populations. The entire genus appears to be an outcrossing system, although the possibility of some inbreeding cannot be excluded. Also there seems to be little hindrance to seed dispersal, even though preliminary studies of resin chemistry may indicate a difference in seed predators in Old and New World species of *Hymenaea.*

Although outcrossing may be frequent, and therefore homozygosity low, it may also be that intense selection results in a high proportion of successful viable recombinations. Each generation may produce a full range of variation through free recombination and high heterozygosity, but selection ensures that the few seedlings which attain maturity, out of a large number that germinate, are phenotypically uniform. However, speciation is occurring in allopatric species with widespread populations and differentiation is occurring in characters which are adaptive or linked to adaptive characters. For example, in the widespread, essentially allopatric species of *H. oblongifolia* and *H. courbaril* there is evidence of constancy in the floral and fruit characters but of differentiation in the foliar characters as populations moved into different habitats. We, therefore, recognized both the unity of the species as diagnosed by reproductive characters, and possible adaptive differentiation in the leaves at the varietal level. This foliar differentiation also occurs in *H. intermedia, H. rubriflora,* and *H. stigonocarpa.*

Sympatric speciation appears to have occurred in the thorn forest of northeastern Brazil and the central Amazonian region. In the thorn forest, *H. eriogyne* and *H. velutina* occur in the same sites but are reproductively isolated by their flowering times. In the central Amazon, *H. parvifolia, H. intermedia,* and *H. reticulata* exist sympatrically and their flowering times overlap. It is presently not known how they remain isolated. It may be, however, a situation similar to that among the Dipterocarpaceae (Ashton, 1969) where a single habitat may be filled with one or more alternative species by chance.

TRENDS OF SPECIALIZATION AND EVOLUTIONARY PERSPECTIVE

Phylogenetic considerations regarding tropical trees must necessarily be speculative because of poor paleobotanical data, generally poor collections, and lack of the kind of experimental work which can be carried out on herbaceous plants. In this study the available morphological, cytological, ecological, and paleobotanical information has been synthesized in an attempt to indicate possible trends in the evolution of *Hymenaea.*

Langenheim (1973) has pointed out that there are few authenticated paleobotanical records for the entire subfamily Caesalpinioideae and that no good evidence for Detarieae (Cynometreae) appears before the Paleocene epoch. The first unequivocal record of *Hymenaea* is found in Oligo-Miocene beds from Mexico (Langenheim, 1966, 1969). However, good tropical megafloral records are poor, due to severe weathering and inaccessibility of outcrops. The most extensive knowledge available has been obtained from pollen. As pointed out previously, *Hymenaea* is probably primarily animal pollinated, which may well be the case for other members of Detarieae – and thus pollen may not get incorporated into the geologic record. Another problem in obtaining fossil evidence for this tribe is that 67% of Caesalpinioideae are endemic to Africa (Brenan, 1965) and relatively few Cretaceous and Tertiary beds occur in Africa. Lack of sediments which might include fossils presents another difficulty in obtaining data (Langenheim, 1973).

Although it is now generally accepted that Africa and South America were once united, geological evidence indicates that drifting of the two continents took place too early (around the Jurassic-Cretaceous boundary) to explain the disjunct distribution of *Hymenaea* or their immediate ancestors. Even if drifting occurred during mid-Cretaceous, as suggested by Veevers, Jones, and Talent (1971) and Raven and Axelrod (1972), it is highly questionable whether any extant angiosperm genera existed at that time. Thorne (1973) also points out that the "tenuous links" between tropical Africa and tropical

America and their stronger floristic relationships to other continents tend to exclude continental drift as a valid explanation for these amphi-Atlantic disjunctions. In a careful analysis of the seed-plant taxa occurring on the two continents, Thorne thinks that the distribution patterns point to occasional long-distance oceanic dispersal rather than retention of ancient widespread ranges which were split by continental drift. Migration, however, was probably easier between the continents during late Cretaceous and early Tertiary, because long after the twc continents were separated there was fiarly direct access across relatively narrow seas and possibly along volcanic islands on mid-ocean ridges. Also, during the Eocene epoch when tropical rain-forest vegetation reached its maximum development, conditions were particularly good for dispersal of disseminules of rain-forest species. With increased population size of the rain-forest species, a greater number of diseminules would be supplied for dispersal and at the same time the area suitable for colonization was larger.

Hymenaea verrucosa in eastern Africa appears to be a relict of a once more continuous distribution in evergreen forests across Africa, and the clear morphological and chemical ties between the African and New World rain-forest species is a significant factor in consideration of the origin of *Hymenaea.* Thus a possible explanation for the origin of the genus and the amphi-Atlantic distribution pattern is an assumed common West African rain-forest ancestral stock (Langenheim and Lee, 1974). From this stock during the early Tertiary, oceanic dispersal to the rain forests of the New World could have occurred – one successful establishment leading to the Cuban species *H. torrei* and another to the development of the Brazilian Atlantic coastal and Amazonian complex of species centering around *H. oblongifolia.* Subsequently the rain-forest species in the New World have provided the stock from which the species in the drier ecosystems have radiated during drying trends in later Tertiary time and during wet-dry oscillations of the Pleistocene.

Paleontologic data suggest that rain-forest vegetation and associated faunas during the Eocene epoch extended as far south as central Patagonia on the South American continent. Menéndez (1969) and Simpson (1969) present evidence from both fossil plants and mammals indicating that strong climatic changes, similar to those in North America and other northern continents, occurred in South America during the Cenozoic. Early in the Cenozoic, climatic zones were less marked, but by the Miocene epoch the climate was on its way to the drier and cooler conditions. That change, with Pleistocene fluctuations, has continued to the present. The Amazon Basin and the adjacent moderate Guayanan and Brazilian highlands (elevations approximately 1000 to 1500 m.) represent the most ancient and stable habitats. Along with the idea of stability of the Amazon Basin has been the common assumption that the Amazonian rain-forest vegetation has probably remained little affected by climatic changes in later Tertiary times and the Pleistocene. In fact, the tremendous diversity of species in equatorial forests has often been explained partially on the basis of climatic stability through time (McArthur, 1969; Baker, 1970). However, recent studies on bird and anolid lizard populations and certain woody plant families have supported the hypothesis that climatic oscillations during the Pleistocene had dramatic effects upon the distribution of the Amazonian *hylaea* (Haffer, 1969; Vanzolini and Williams, 1970; Vuilleumier, 1971; Vanzolini, 1973; Prance, 1974). They further postulated that during dry phases the rain forest only occupied disjunct areas which served as refugia for rain-forest organisms. Over extended periods of time, local selective pressure could have produced differentiation among populations of species inhabiting different forest refugia.

In humid phases, rain-forest vegetation re-expanded and, where previously isolated populations reunited, formed a complex pattern of speciation.

The correlations in time of the humid-arid cycles with South American glacial events remains controversial and there is presently little geographical or paleobotanical evidence to support this hypothesis. However, there is convincing geological evidence for these cycles from Africa (Moreau, 1966) and future geological work may provide the necessary data. Paleontological data from northern South America (Van der Hammen, 1961) indicate repeated vegetational changes in the Pleistocene over large areas. Also the present floristic diversity in the Amazonian *hylaea*, with relatively large patches of savannahs characterized by *hylaea* vegetation, suggest the influence of drier climates in the past. *Hymenaea* appears to be an example of a genus which probably had its origin and early history in rain-forest conditions in Africa and South America but which then differentiated in response to dry conditions within the *hylaea* as well as over a wide range of drier habitats both north and south of the *hylaea.*

Although *H. oblongifolia* appears to be the New World parental stock, there is little evidence to indicate whether the African ancestor arrived only in the Brazilian "bulge" area or in the more northerly Amazonian region. The primary distribution of the species today is represented by variety *oblongifolia* in the *várzeas* of the high annual rainfall areas of the eastern and western zones of the Amazon Basin and it is conspicuously absent from the drier central areas. Schultes (1953) has indicated that variety *oblongifolia* in the western zone is associated with the ancient Cretaceous outliers of the Venezuela-Guyana landmass. The variety *palustris* has a similar distribution to the typical variety but has been differentiated in the *igapó* habitat. One collection of variety *palustris* on the western side of the Andes is an additional indication of early Tertiary occupation of this area, *i.e.,* prior to the uplift of the Andes. Cuatracasas (1957) also has pointed out that the *Choco* flora is high in endemics and probably represents a very old extension of the Amazonian flora with which it shows strong affinities. He states that the region was probably isolated from the Amazon Basin in the Pliocene, resulting in the differentiation of many species. The variety *latifolia* is presently restricted to the *Hylaea Bahiana* where it occurs in upland sites (although commonly found along the streams). Morphologically (the petals being especially noteworthy), variety *latifolia* is more closely related to *H. verrucosa* than variety *oblongifolia.* This similarity possibly indicates that variety *latifolia* represents a more primitive stock than the more widely distributed typical one in the Amazonian region. Variety *davisii* has only been found in moist habitats along rivers in the *hylaea* in Guyana, where it differentiated again under different conditions from the other varieties. Thus, from the range of habitats in which varieties of *H. oblongifolia* occur, it appears to be a plastic stock in which some morphological differentiation (leaves) has taken place, parallel with ecological isolation.

Hymenaea parvifolia appears most closely related to the *H. oblongifolia* complex, but represents a drier habitat variant, which occurs on sandy *terra firme* sites within the lowest rainfall belt in the eastern and central part of the Amazon Basin. Considerable morphological variability characterizes populations of *H. parvifolia,* a variability also reflected in the quantitative composition of leaf pocket resins. *Hymenaea parvifolia* occurs in the *campo* and savannah areas more than any other species of *Hymenaea* in the *hylaea.* In form it ranges from shrublike in the center of the *campo* to an emergent tree in the neighboring rain forest. Apparently non-*hylaea* species characterize most *campos;* however, *H. parvifolia* occurs in both habitats. It is in these *campo-hylaea* contact areas where *H. par-*

vifolia displays its greatest variability, thus possibly providing evidence supporting Haffer (1969), Vanzolini (1973), and Prance (1974).

Although *H. aurea* occurs in the same area of the Atlantic coastal forest as *H. oblongifolia* var. *latifolia,* the two rarely occur together and *H. aurea* also has been reported further south in the forests around Rio de Janeiro. It may have evolved from *H. oblongifolia* stock, as might *H. eriogyne* and *H. maranhensis,* both of which appear to be relatively closely related to *H. aurea.* In fact, these three species form a morphologically coherent group. *H. eriogyne* is a low tree restricted to thickets or "forest patches" over a relatively wide area of the *caatinga* of northeastern Brazil. Its morphological characteristics (including type of seedling) indicate its relationships to the rain-forest species, and suggests that it is a relict from the rain-forest vegetation that probably once covered this part of Brazil. *Hymenaea maranhensis* has been found only in southeastern Maranhão where it occurs in a type of *cerrado* locally called "chapada." It has several characters which relate it to some of the primitive rain-forest species. These, together with the proximity of its range to the Amazon Basin, suggest that it might be a relict. However, several presumably adaptive characters, such as the relatively large and persistent bracteoles, the thickly coriaceous leaves, and the shrubby habit suggest an evolutionary trend toward a drier habitat. Its resemblance to the more common *cerrado* species, *H. stigonocarpa,* also suggests an intermediate position between that species and those of the Amazonian *hylaea.*

Hymenaea courbaril, with its six varieties, occurs in every major tropical ecosystem type, thus giving evidence of remarkable ability to adjust to a gradient of moisture conditions. *Hymenaea courbaril* var. *courbaril* seems to be less common in the Amazon Basin, where it appears relatively restricted to *terra firme* sites in the eastern zone, than in northern South America and Meso-America, where it is common in most ecosystem types and may even attain dominance in certain dry forest ecosystems. It also occurs south of the Amazon Basin but is not as important there as in its northern range. However, in the Amazon and south of it, *H. courbaril* is represented by several varieties and it also appears to have been the parental stock for the radiation of several *cerrado* species. In the Amazon Basin, *H. courbaril* seems best represented by its variety *subsessilis* which is most common along the sandy beaches of the tributaries (although it can occur generally on the *terra firme*) in the central Amazon region. Ducke (1935) reported that specimens from the *campos* near Almeirim, Pará, appear to be intermediate between variety *subsessilis* and the typical one. Variety *subsessilis* along the beaches is much reduced in size, essentially becomes denuded of leaves during the dry season, and flowers on a different schedule from variety *courbaril.* It could be a variant which developed during a dry interval of the Pleistocene. Variety *stilbocarpa* until recently was considered a separate species, although Ducke (1935) had earlier suggested that it probably was a subtropical variety of *H. courbaril.* It appears to have radiated into a distinctive habitat, essentially coincident with the *cerrado,* although it occurs as a component of forests along river valleys. The typical variety of *courbaril* can occur in the same sites but is much less common than variety *stilbocarpa. Hymenaea courbaril* var. *altissima* is another subtropical variety with a narrow distribution along the Atlantic coast.

Hymenaea courbaril var. *longifolia* has been collected in *cerrado* in the highlands of the Chapada do Araripe on the border of Ceará, Piauí and Pernambuco, surrounded by *caatinga.* The difference in vegetation on the highland is probably due to more rainfall than in the areas below. This variety seems closely related to both *H. velutina* and *H. stig-*

onocarpa and may possibly have been the stock from which both species were differentiated. The latter is a widely spread *cerrado* (and to a lesser extent *caatinga*) species which has differentiated sufficiently to permit recognition of three varieties. *Hymenaea stigonocarpa* var. *stigonocarpa* is particularly close to *H. courbaril* var. *longifolia.* The variety *pubescens* is more widespread than either varieties *stigonocarpa* or *brevipetiolata. Hymenaea velutina* is a small tree restricted to thickets on rock outcrops in the *caatinga* of northeastern Brazil, frequently occurring sympatrically with *H. eriogyne.*

Hymenaea intermedia is morphologically intermediate between *H. oblongifolia* and *H. courbaril. Hymenaea intermedia* generally occurs on *terra firme* in a relatively restricted area of the drier parts of the eastern and central Amazon. Another variety, *adenotricha,* which only recently was reduced to this status (Lee and Langenheim, 1974a), occurs in the western Amazon. This suggests that the range of the taxon was probably more extensive than previously thought. Ecologically, *H. intermedia* is closer to *H. courbaril* than to *H. oblongifolia.* It seems possible that *H. intermedia* represents the stock most closely related to *H. oblongifolia* from which the physiologically plastic *H. courbaril* could have evolved. The evidence for *H. courbaril* or *H. intermedia* occurring as remains in amber, the chemical relationships of the amber itself from Chiapas, Mexico (Langenheim, 1966) and the present distribution of *H. courbaril,* lead us to believe that *H. courbaril* (and some of its closely related species in dry ecosystems) evolved during the drying trends in the mid-Tertiary.

Hymenaea reticulata, although a distinctive species, has characters similar to *H. rubriflora* and *H. courbaril.* Until recently, it appeared to have a restricted distribution in relatively dry *terra firme* sites in the central Amazon near Manaus. However, a specimen, which is now considered to be *H. reticulata,* from Iquitos, Peru may indicate a wider distribution in the past than previous collections suggested.

The diagnostic characters of *H. rubriflora* are similar to those of *H. reticulata,* which suggests their relationship. *Hymenaea rubriflora* occurs in the Atlantic coastal forests from Rio Grande do Norte southward to Espirito Santo and Rio de Janeiro. Thus it has the widest distribution known for any *Hymenaea* restricted to the Atlantic coastal humid forests. The distribution of *H. rubriflora* and *H. reticulata* is highly disjunct, which may be explained by an early separation and subsequent speciation from a once more widely distributed ancestral species.

Careful analysis of morphological characters and those that affect the reproductive biology in living *Hymenaea* species appears to substantiate the hypothesized historical events. In working out the possible phylogeny of *Hymenaea,* we have attempted to assess evolutionary trends shown by certain morphological characters. A discussion of these trends follows.

One of the major differences separating the two sections is the basic type of inflorescence. The long panicle which characterizes sect. *Trachylobium* is assumed to be the more primitive form with the corymbose inflorescence in sect. *Hymenaea* being derived. The general acceptance of the panicle as a more primitive inflorescence than the corymb (Nägeli, 1884; Celakovský, 1892; Pilger, 1922; Goebel, 1931; Rickett, 1944) agrees with the earlier assumption that *Hymenaea* has been derived from a common ancestral stock in Africa (Langenheim and Lee, 1974), because *H. verrucosa* and its neotropical allies, *H. oblongifolia* and *H. parvifolia,* are characterized by a paniculate inflorescence.

Several characters also appear to result from the transition of panicle to corymbose inflorescence, such as (1) shortening of inflorescence branches, (2) reduction of number of flowers in individual inflorescences, and (3) increase in size of flower. Some floral characters have been found to be highly correlated with type of inflorescence. Thus, their relative primitive or advanced status might be deduced.

Several quantitative characters, such as size of fruit, and number of ovules per ovary (hence, number of seeds per fruit) are highly correlated with number of flowers per inflorescence and size of flower. In paniculate inflorescences, fruit are often rather small, ovate to obovate, and few-seeded, whereas the larger, oblong to cylindrical, and many-seeded fruits are formed by flowers in corymbose inflorescences. Stebbins (1971) points out that number of seeds produced by an individual plant is determined by (1) number of seeds (ovules) per ovary, (2) number of flowers per inflorescence, and (3) number of inflorescences. Reduction in number of fruits produced by corymbose inflorescences appears to be compensated for by the larger number of seeds produced per fruit. Selection for increased seed number may bring about profound alteration of the gynoecium and other flower parts (Stebbins, 1971).

The presence or absence of a claw in a single petal might be interpreted as either primitive or advanced. For example, Cowan (1953) faced this dilemma in the genus *Macrolobium.* He first thought that clawed petals were more advanced, having originated by elongation of a basal portion of the blade. If this were assumed to be correct, the highly evolved flower of *Macrolobium* sect. *Stenosolen* would possess the more primitive petals. Therefore, considering the totality of characters, Cowan suggested an alternate hypothesis that the subsessile or sessile petals characteristic of sect. *Stenosolen* were probably evolved by progressive shortening of the claws. Similarly in *Hymenaea,* when clawed petals are considered in relation to other characters, they appear to be a primitive trait. Supporting evidence might also be provided by *Eperua,* a New World genus related to *Hymenaea.* The flowers of *Eperua* are characterized by a single large, broad, mostly sessile petal. This condition appears to be derived, especially in view of its highly specialized inflorescence.

The pattern of vesture on the ovary also provides characters suggesting a similar evolutionary trend. In presumably more primitive species, such as *H. oblongifolia* and *H. parvifolia,* the ovary is usually pubescent throughout and densely hirsute at the base. This type of pubescence is also found in *H. eriogyne,* a relict in the *caatinga,* and *H. aurea,* restricted to the Atlantic coastal forests. Both of these species are thought to have evolved in the early rain forests and have subsequently been restricted in their distribution. In the remainder of the species there appears to be a trend toward reduction in the pubescence on the ovary. In *H. verrucosa,* one of the primitive species thought to be relictual in eastern Africa now, the ovary is glabrate except for the basal hairs. In *H. maranhensis,* the hairs are restricted to the base and the margin of the ovary, and in *H. martiana* it is glabrate but there always is a tuft of basal hairs. Even on the glabrate ovary found in the remaining species, scattered hairs can still be found. Thus it appears that the glabrate ovary is probably derived from a pubsescent one.

Flowers of *H. courbaril, H. martiana, H. velutina,* and *H. stignonocarpa* have a relatively large, elevated disc with an additional mass of nectariferous tissue shaped like an inverted bell inside the hypanthium (such tissue may attain a volume up to one cubic centimeter in *H. stigonocarpa* vars. *stigonocarpa* and *pubescens*). Smaller and less well-defined nectariferous tissue is found in most of the species in sect. *Hymenaea.* However, a disc without any apparent mass of nectariferous tissue occurs in the three species in sect. *Trachylobium.* Thus a well-developed nectariferous disc appears to be a specialized structure.

It seems possible that the numerous correlations of "primitive" inflorescence and floral characters found in the two sections suggest evolutionary trends responding to new habitats and new pollination vectors. Perhaps, as indicated by Stebbins (1970), one of the most important features of floral-character syndromes is that they favor correlations between characters and the recognized behavior of predominant pollinators. In *Hymenaea,* one of the most conspicuous of these is between paniculate and corymbose inflorescences. The long panicle characteristic of sect. *Trachylobium* would provide a "landing platform" upon which insects could alight before entering the flower to seek food. The arrangement of the numerous small flowers on the periphery of a conical inflorescence supported by slender branches is best suited for small pollinators such as bees. The robust flower and rather short branches of the corymbose inflorescence found in sect. *Hymenaea,* on the other hand, are better suited for visitation by larger animals. This inflorescence together with such other characters as development of massive secretory tissue and production of more abundant nectar would also favor pollination by larger vectors such as bats and moths. This agrees with Stebbins' view that evolutionary shift from one vector to another is probably triggered by entrance of the plant into a habitat where the original vector is scarce and a new vector is abundant.

The African *H. verrucosa* is characterized by a coarsely verrucose-rugose fruit. Among the New World species all but the Cuban *H. torrei* have a smooth fruit surface. Occurrence of verrucose fruit in *H. torrei,* and similar but smaller resinous pockets on fruit of some *Hymenaea* species, suggests the possibility that verrucose fruit represents a more primitive condition. However, since the resin chemistry of the diterpene acids is very different in the verrucose fruits of *H. verrucosa,* compared with smooth fruits of *H. oblongifolia,* it might be assumed that there were different seed-predator pressures in South America than in Africa.

The pollen grains of various species of *Hymenaea* thus far investigated belong to the same basic pattern, although two extreme conditions in the fine structure of the exine occur (Whittingham, unpubl. ms.). In *H. verrucosa* and *H. oblongifolia* the exine is characterized by minute puncta which occur obscurely in linear arrangement between series of ridges. In the species belonging to section *Hymenaea,* however, the puncta of the exine are conspicuously larger and are variously arranged either diffusely, or in clusters or rows. Some species appear to be somewhat intermediate toward the condition seen in *H. oblongifolia,* indicating the possibility of a transition. Furthermore, species whose pollen have the thickest and most complex exines are members of sect. *Hymenaea.*

Hymenaea seedlings fall into subtypes depending upon length of the epicotyl. It is evident that species now found in wet or moist habitats, or those that are relicts of such ecosystems, are characterized by a very short or almost sessile epicotyl, whereas those with a distinct epicotyl are mostly inhabitants of drier ecosystems. The epicotyl is particularly useful when correlated with other morphological characters, such as type of ovary pubescence and of inflorescence. Based on comparative study of these features, it is concluded that species with short or sessile epicotyls are either direct descendants of rain-forest species or relicts and those with long epicotyls result from radiation of rain-forest elements into drier habitats. Although no selective advantage is evident, a long epicotyl appears to be derived from a sessile one.

A dorsal median groove on the cotyledon of *Hymenaea* seed may be of evolutionary interest. Such grooves have been found in several presumably more primitive species but are absent in more advanced ones.

The nature of bracteoles in *Hymenaea* possibly indicates adaptive trends toward a drier habitat. In most species, the two bracteoles are relatively small and early-deciduous. On the other hand, in *H. stigonocarpa* and *H. maranhensis* they persist through most of the bud stage or, in the latter species, even into anthesis, thus providing a protective envelope for the developing flower.

In summary, possible morphological trends of development in the evolutionary history of the genus of *Hymenaea* are tabulated as follows:

Paniculate to corymbose inflorescence
Many-flowered to few-flowered inflorescence
Small to large flower
Long, slender, and slightly flexuous to short, stout, and strongly flexuous inflorescence branches
Clawed to sessile petals
Few-ovuled to many-ovuled ovary
Pubescent to glabrous ovary
Absence of additional mass of nectariferous tissue to its presence within the hypanthium
Exine of pollen having minute puncta to much larger puncta
Ovoid to oblong fruit
Few-seeded to many-seeded fruit
Small to larger fruit
Verrucose to smooth fruit
Cotyledon with dorsal median groove to its absence
Early-deciduous to persistent bracteoles
Seedling with short to long epicotyl

Figure 10 shows our present view of the specific interrelationships and indicates radiation from rain forest into a variety of dry ecosystem types. A line in the diagram from one species to another indicates their close relationship only and does not necessarily imply that one has given rise to another.

ETHNOBOTANY

One of the commercially valuable products of *Hymenaea* is the pale yellow or reddish resin, sometimes referred to as a "gum-resin." This resin produced in Brazil has been known in the trade as "South American Copal" to distinguish it from various copals from other parts of the tropics (Tschirch and Stock, 1936). Throughout the wide range of the genus in the New World, natives use the resin for incense and generally as a cement. Extensive commercial usage, however, seems to have developed primarily in Brazil, where it has been used in manufacture of special grades of varnish and to a limited extent for medicinal purposes (Langenheim, 1967). Howes (1949) reported that Brazilian copal is recovered frequently from water courses and varies from the size of a pea to that of a man's fist. When dug from the ground, it may be in large, irregularly shaped pieces or blocks which occasionally weigh as much as 3 kg. (Ducke, 1925). Record and Hess (1943) noted that native collectors may obtain a barrel of copal by digging around the roots of a tree, and several barrels may be secured by digging what was doubtless the former site of a large tree, long since decayed. The source of this resin previously has been thought to be primarily *H. courbaril* (Howes, 1949; Ducke, 1935; Langenheim, 1967), although other species of *Hymenaea* may be responsible. Even though resin apparently has been obtained from various

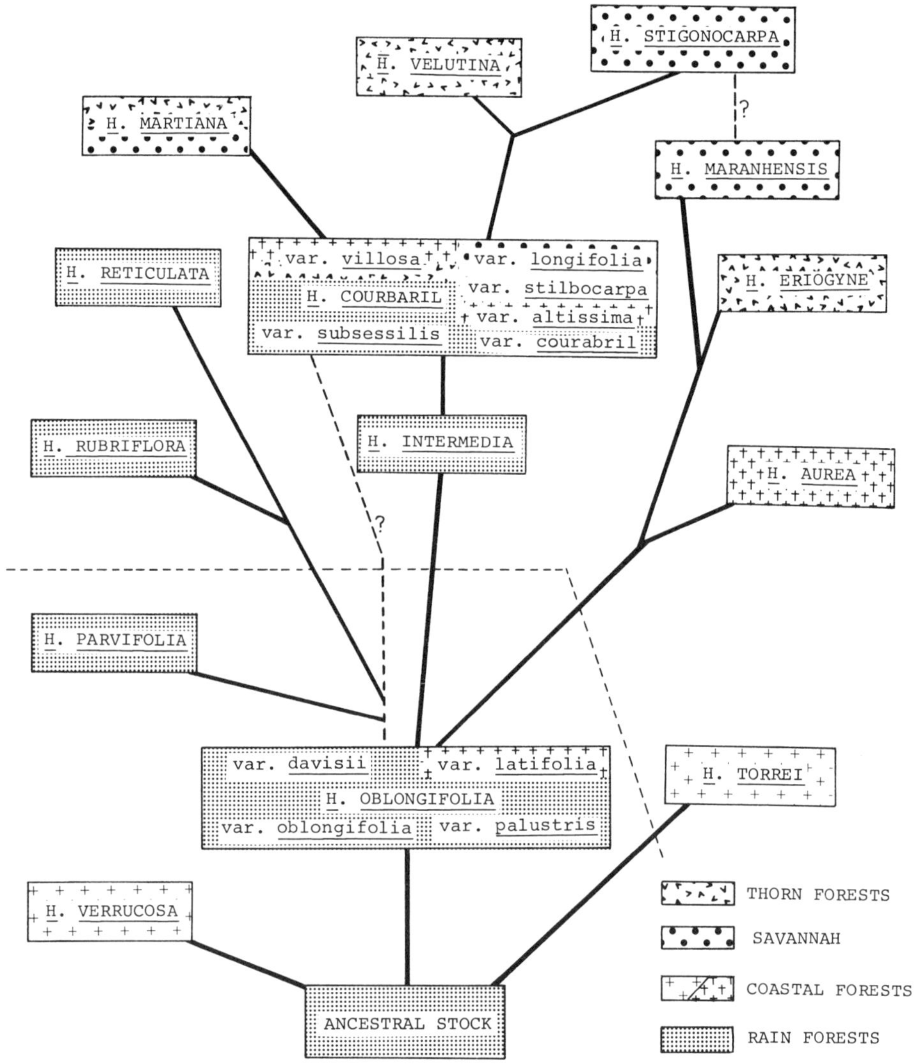

10. Diagram of putative relationships within *Hymenaea.*

sources in Brazil, the greatest collecting center seems to have been in the Amazonian region – the Ilha do Marajó near Belém and along such Amazonian tributaries as Tapajós and Xingu (Langenheim, 1967). Because *H. courbaril* is scarce in these areas, it seems likely that other more common species such as *H. oblongifolia* and *H. parvifolia* might be responsible for the Brazilian resin used commercially (Langenheim, Lee, and Martin, 1973). This view is supported by the chemistry of the resin.

Noriega (1918) indicated that large masses of resin found in the soil in Oaxaca, Mexico were of economic value. Pre-Colombian Indians in southern Costa Rica and Panama carved figurines from *Hymenaea* resin (Balser, 1960; Langenheim, unpubl. data). Oligo-Miocene amber, derived from *Hymenaea* from Chiapas, Mexico is still being used for jewelry as well as incense (Langenheim, 1966; 1969). Other smaller deposits of *Hymenaea* amber occur in Miocene beds near Medellín, Colombia and Pliocene strata from Pará, Brazil.

In East Africa, *H. verrucosa* is known as the source of Zanzibar Copal, which is also known in the English market as "Anime." According to Kirk (1868) this was the most valuable of all resins for manufacture of varnish exceeding any produced on the West African coast for hardness, elasticity, and polish. Zanzibar Copals were gathered in a relatively large quantity and were highly valued at that time.

On both continents resin can be collected from living trees, from the soil in a "semi-fossilized" form, having dropped from living trees, or "fossilized" from areas where trees no longer exist. The longer the resin is buried in the soil the better its quality and the harder it is likely to be. Resin collected directly from living trees is the poorest grade. Le Cointe (1947) reports that resin from *H. parvifolia* is dissolved in xylol and used in preparation of microscopic slides.

The wood of *Hymenaea* is of good quality and is used for carpentry and construction of all kinds, such as ship and boat building, furniture, cabinet work, etc. (Record and Hess, 1943). It was formerly well-known in European markets but is less common now, perhaps because of its relatively slow growth and the inadequate supply.

The bark has been used by Amazonian Indians for making canoes. The process, according to Record and Mell (1924) and Lima (1950), consists of removing the bark in one large piece, preferably from a standing tree, sewing it together at the ends with bush rope or bark-fiber cordage, waterproofing the seams with resin, and then fitting a wooden cross-piece to hold the shape. The bark of old trees is often over an inch thick, and canoes with a carrying capacity of twenty-five to thirty men can be made from a single tree. A similar type of canoe also has been used in Guyana (Sandwith, 1938).

Hymenaea leaves, particularly those of *H. courbaril,* are used regionally by native people to make tea; in northeastern Brazil small pieces of bark, usually soaked in Cochaça (Brazilian rum), are believed to have medicinal properties.

The pulpy tissue surrounding the seeds of all *Hymenaea* species is edible and fruits are sold in local markets. The seeds of *H. parvifolia* also can be used to prepare a latex (Mors, 1946).

TAXONOMIC TREATMENT

HYMENAEA L. SP. PL. 1142, 1753 (GEN. PL. 1ST ED. 366, 1737).

Courbaril Piso, Hist. Nat. Bras. 123. 1658.
Courbari Adanson, Fam. Pl. 2: 317. 1763.
Hemenaea Scop. Introd. Hist. Nat. 296. 1777.
Tanroujou Juss. Gen. Pl. 578. 1789.
Trachylobium Hayne, Flora 10: 743. 1827.
Hymenia Griff., Icon. Pl. Asiat. 4. 1854.

Unarmed resinous trees, occasionally shrubby, 1.5-40 m.(+) tall, evergreen to deciduous. Leaves bifoliolate, pellucid-punctate, petiolate, and stipulate, the stipules 2, early-deciduous, the leaflets with twisted petiolule, the lamina ovate, obovate, oblong, or falcate, the inner half much narrower than the outer half, coriaceous. Inflorescence long- or short-paniculate and becoming densely corymbose during anthesis, the bracteoles 2, early-deciduous or rarely persistent. Flowers pedicellate, all parts pellucid-punctate; hypanthium campanulate, disc-bearing, with additional mass of nectariferous tissue within or not; calyx lobes 4, strongly imbricate in bud, ovate to obovate; petals 5, subequal, creamy white to rose-tinged, rarely red, sessile to distinctly unguiculate; stamens 10, exserted, the filaments folded in bud, the anthers dorsifixed and versatile; ovary stipitate, occasionally subsessile (the stipe laterally

adnate to the hypanthium wall), the ovules 3-16(+). Fruit ovoid to obovoid, rhomboidal, or oblong, compressed or not, woody, indehiscent, the surface coarsely verrucose-rugose to smooth, dotted with resinous pockets and minute white to pink lenticels. Seed(s) 1-8(+) surrounded by endocarp, ovoid to obovoid or ellipsoid, the testa light to dark brown, its intrusion into cotyledons producing 2 or 4 shallow furrows.
Type species: *Hymenaea courbaril* L.

KEY TO THE SECTIONS OF HYMENAEA

1. Inflorescence long-paniculate, 18-35 cm. long or longer, its branches long, slender, only slightly flexuous; flowers small (older buds usually 10-15 x 6-8 mm.*), ovary usually pubescent throughout, densely hirsute at base; fruit ovoid to obovoid, subcompressed, 1- to few-seeded. . . . . Sect. 1. *Trachylobium*
1. Young inflorescence short-paniculate, 8-15 cm. long, corymbose when mature, its branches short, thick, strongly flexuous; flowers large (older buds 12-35 x 6-18 mm.); ovary densely woolly-tomentose to glabrate; fruit obovoid, rhomboidal but mostly cylindrical to oblong, compressed or not, usually many-seeded . . . . Sect. 2. *Hymenaea*

Section 1. *Trachylobium* (Hayne) Baill.

Type Species: *Hymenaea verrucosa* Gaertn.

Inflorescence long-paniculate, of distinctly terminal and lateral racemes up to 18 cm. long, its branches rather long (to 5 cm.), slender and slightly flexuous. Flowers small, disc-bearing but lacking any apparent mass of nectariferous tissue, the ovary densely hirsute at base, densely pubescent or ± glabrate above. Fruit rarely exceeding 6 x 3 x 2.5 cm.,** smooth or coarsely verrucose-rugose. Seeds 1 or 2, rarely more.

KEY TO THE SPECIES OF SECTION 1. TRACHYLOBIUM

I. Fruit coarsely verrucose-rugose; ovary glabrate except at base . . . . 1. *H. verrucosa*
I. Fruit smooth; ovary pubescent throughout.
  II. Petals densely pilose above; leaflets small, 5-8(12) x 2.5-3.5(4.5) cm., broadly falcate . . . . 3. *H. parvifolia*
  II. Petals glabrous; leaflets large, 11-15 x 4-6 cm., oblong to falcate . . . . 2. *H. oblongifolia*

Section 2. *Hymenaea*

Courbaril Baill. nomen, Hist. Pl. 2: 114. 1870.
Type species: *Hymenaea courbaril* L.

Young inflorescence short-paniculate, densely corymbose at maturity, of rather short terminal and lateral racemes, its branches short, stout, and strongly flexuous. Flowers large, disc-bearing, with additional mass of nectariferous tissue or not, the ovary glabrous throughout to densely woolly tomentose, rarely densely hirsute at base. Fruit large, smooth (coarsely verrucose-rugose in *H. torrei*), oblong to cylindrical, subcompressed to strongly compressed, occasionally rhomboidal. Seeds 3-8 or more, rarely only 1 or 2.

KEY TO THE SPECIES OF SECTION 2. HYMENAEA

I. Surface of fruit coarsely verrucose-rugose . . . . 4. *H. torrei*

---

*The dimensions of the flower are taken from buds prior to opening. Length is from base to apex, diameter follows, measured at the widest part.

**In this treatment, the dimensions of fruit are given in following order: length x width x thickness. In all cases the maximum dimensions are recorded. In an ovoid fruit, the width is measured from the side of the suture to the opposite side, whereas the thickness is measured at right angles to the width. The dimensions of seeds are expressed similarly.

I. Surface of fruit smooth, usually with prominent small resinous pockets or such structures obscure.
  II. Ovary mostly densely pubescent throughout; petals spatulate, distinctly unguiculate, the claws more than 6 mm. long.
    III. Ovary densely pubescent throughout; bracteoles early-deciduous.
      IV. Leaflets obovate, pubescent beneath or occasionally glabrous; ovary moderately hirsute at base; fruit narrowly oblong, subcompressed . . . . 5. *H. eriogyne*
      IV. Leaflets obovate-oblong, densely golden brown tomentose beneath and sparsely tomentose above; ovary densely wooly tomentose throughout; fruit broad and flat . . . . 6. *H. aurea*
    III. Ovary densely woolly tomentose at base and marginally, otherwise glabrate; bracteoles occasionally persistent throughout anthesis. . . . . 7. *H. maranhensis*
  II. Ovary glabrate throughout or sparsely pilose at base; petals ovate, oblong, or oblanceolate, not spatulate, sessile to subclawed, the claws less than 2 mm. long.
    III. Ovary borne on a slender stipe, rather small (2 x 2 mm.), rhomboidal; fruit ovoid or obovoid, rhomboidal at maturity, subcompressed; seeds 1 or 2, rarely more . . . . 10. *H. intermedia*
    III. Ovary rigidly stipitate, occasionally subsessile, oblong-oblique (6 x 2.5 mm.); fruit oblong to cyclindrical; terete to strongly compressed; seeds 3-6 or more.
      IV. Petals narrowly oblanceolate, acute, slightly unguiculate, the claws *ca.* 2 mm. long.
        V. Leaflets falcate, glabrous, dull, the veinlets elevated, prominantly reticulate; fruit large, rhomboidal-oblong (10-20 x 6-10 x 2.5-3.5 cm.), broad, smooth (without evident resinous pockets and lenticels), feltlike. . . 8. *H. reticulata*
        V. Leaflets oblong, sparsely pubescent, shining above, densely golden brown tomentose beneath, or glabrous and shining on both surfaces, the veinlets obscure; fruit small (5-6 x 3 x 1.5-2 cm.), oblong, with evident resinous pockets and lenticels, not feltlike. . . . . 9. *H. rubriflora*
      IV. Petals ovate to obovate, sessile to subsessile, rarely slightly unguiculate.
        V. Ovary usually pilose basally by a tuft of hairs, occasionally glabrate throughout; leaflets ovate to obovate, glabrate to sparsely pubescent above, densely greyishwhite tomentose beneath . . . . 12. *H. martiana*
        V. Ovary glabrate throughout (very rarely sparsely pilose at base); leaflets falcate, ovate, obovate to rounded, glabrous to densely pubescent.
          VI. Leaflets glabrate throughout, rarely villous beneath; flowers medium-sized (older buds** 12-32 x 6-12 mm.); hypanthium campanulate, with stalklike base *ca.* 6-8 (+) mm. long.
            VII. Leaflets mostly falcate, occasionally obovate-oblong, short-acuminate to obtuse; base slightly to not at all prolonged on one side; fruit medium-sized to large (largest *ca.* 20 x 9 x 4.5 cm.), moderately compressed, usually dotted with small, prominent, resinous pockets. . . . . 11. *H. courbaril*
            VII. Leaflets obovate to oblong, obtuse or retuse; base extremely prolonged on one side; fruit medium-sized (*ca.* 8-11 x 4-5 x 3-4 cm.), slightly compressed, with wide faces marked by ± conspicuous, subdiagonal veins, smooth and ± feltlike . . . . 13. *H. velutina*
          VI. Leaflets glabrate, or more often densely pubescent beneath, medium-sized to large (18-24 x 10-16 (+) cm.); flowers large (older buds *ca.* 35 x 18 mm.); hypanthium broadly obconical, subsessile or with stalklike base *ca.* 2 mm. long. 14. *H. stigonocarpa*

**The dimension of the flower are taken from buds prior to opening. Length is from base to apex, diameter follows measured at the widest part.

1. *Hymenaea verrucosa* Gaertn. Fruct. 2: 206, pl. 139, fig. 7. 1791 (Figure 11)

Trachylobium verrucosum Oliver, Fl. Trop. Afr. 2: 311. 1871.

T. hornemannianum Hayne, Arzneik. gebräuchl. Gewächses 11: pl. 15. 1830; *Hymenaea hornemanniana* D. Dietr. Syn. Pl. 2: 1473. 1840. Based upon *Hornemann,* Mauritius (C, holotype!).

Trachylobium gaertnerianum Hayne, Arzneik, gebräuchl. Gewächses 11: pl. 19-a. 1830; *Hymenaea gaertneriana* D. Dietr. Syn. Pl. 2: 1473. 1840. Based upon *Martius,* Java (P?, holotype).

Trachylobium lamarckianum Hayne, Arzneik. gebräuchl. Gewächses 11: pl. 19-b. 1830; *Hymenaea lamarckiana* D. Dietr. Syn. Pl. 2: 1473. 1840. Based upon *Lamarck,* Madagascar (?) (C, holotype!).

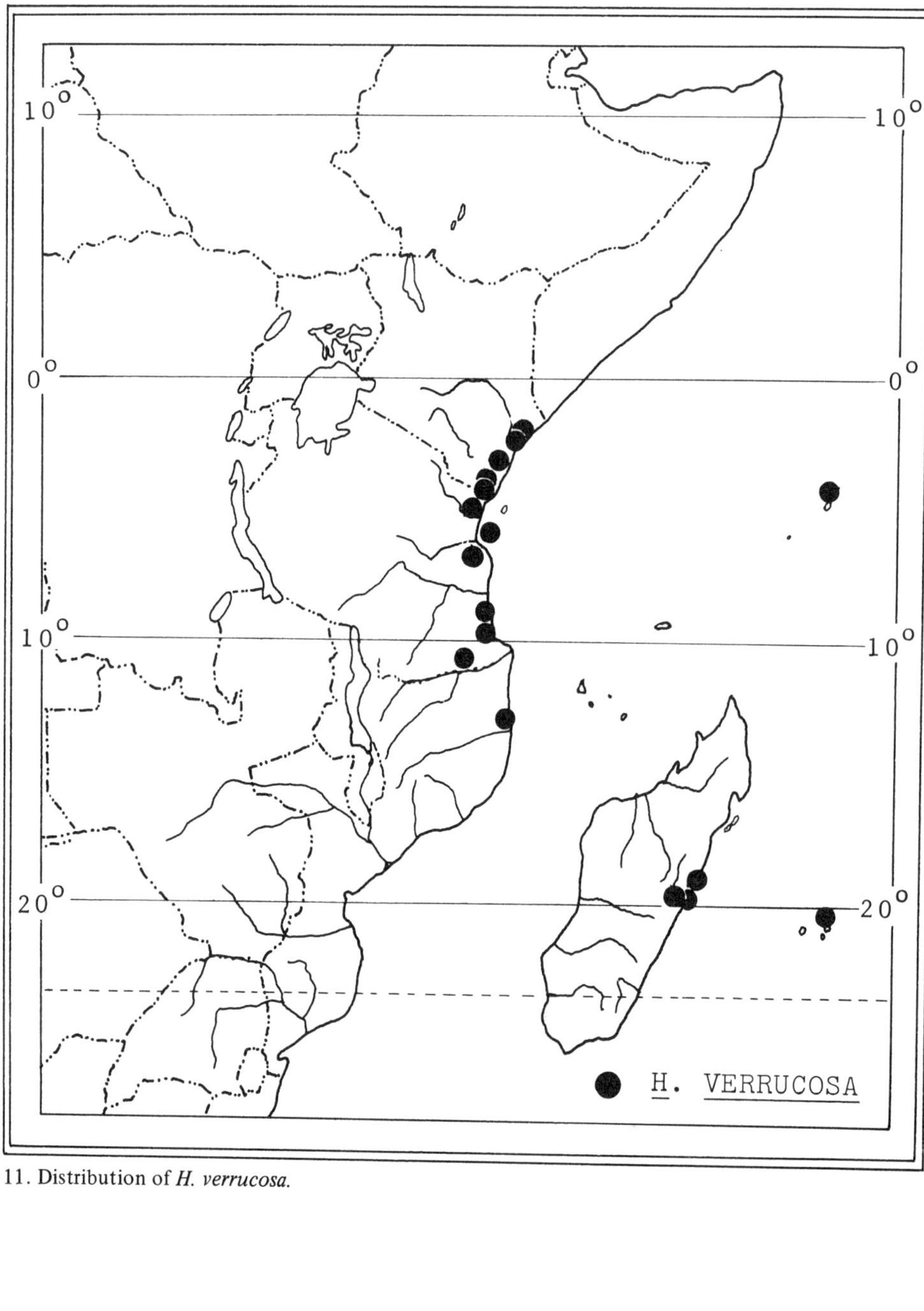

11. Distribution of *H. verrucosa.*

Trachylobium mossambicense Klotzsch in Peters, Reise Mossamb. Bot. 1: 21, pl. 2. 1861. Based upon *Peters,* Mozambique, Querimba (Quisanga) (B, holotype; K!, isotype).

Evergreen tree 6-24 (-40) m. tall, the young branchlets puberulent to glabrate. Leaves petiolate, the petioles 8-18 mm. long, the stipules narrowly elliptic, *ca.* 20 x 4-6 mm., obtuse, caducous, the leaflets with twisted petiolule 3-5 mm. long, the lamina elliptic to elliptic-oblong, broadly falcate, 3.5-12 x 2-6 cm., coriaceous, glabrate and shining on both surfaces but duller beneath, obtuse to short-acuminate, the base oblique, with the inner side acute and the outer side moderately rounded. Inflorescence paniculate, the panicle up to 35 x 25 cm., often much smaller and occasionally few-flowered, its branches slender and ochraceous-puberulent, the flowers spirally arranged along ultimate branches, the pedicels articulate at base, 2-4 mm. long, bracts and bracteoles ovate to orbicular, caducous. Flowers small; hypanthium campanulate, subsessile; calyx lobes ovate, 7-11 x 4-6 mm., densely appressed-pubescent outside, silvery-sericeous inside; petals 5, either (a) the 3 upper clawed, suborbicular, and subequal and the lower 2 minute and scalelike, or (b) all clawed and subequal, the large petals 15-20 mm. long, their lamina 8-10 mm. in diameter; stamens *ca.* 25 mm. long; ovary slightly stipitate, densely hirsute at base, ± glabrate above; ovules 4. Fruit ovoid or obovoid to oblong-ellipsoid, 2.5-5.0 x 1.5-3.0 x 1.3-2.5 cm., coarsely verrucose-rugose, reddish to dark brown. Seeds 1-3, ellipsoid, 13-18 x 9-12 x 8 mm., the testa dark brown, its intrusion into the cotyledons producing 2 longitudinal and 2 vertical, shallow furrows. Chromosome number 2N = 24.

Type: Madagascar, *Gaertner* (TUB – holotype!).

Habitat and distribution: On or near coast in lowland dry evergreen forest, woodland, and coastal evergreen bushland. This species is restricted to the coast of East Africa and the islands of Zanzibar, Seychelles, and Mauritius and the east coast of Madagascar. It flowers from February to April. *Hymenaea verrucosa* is commonly planted in tropical gardens (*e.g.,* in Java, Ceylon, India, Hawaii, Tahiti, Canal Zone, Cuba, etc.).

Specimens examined: KENYA. Kilifi District: Arabuko, III-IV-1929, *R. M. Graham B263* (A, K, US); Coastal Prov., 22-XI-1969, *R. E. Perdue & S. P. Kibuwa 10045* (K); Gedi, 30-IV-1954, *E. C. Trump 100* (K); Nairobi District: 21-I-1964, *B. V. Schmidt 3955E;* Kwale Reserve 11-VII-1968, *Langenheim 5520* (UC); Jilore Forest Station, 13-VII-1968, *5522* (UC).

MADAGASCAR. Between Tamatave & Antananariva, VII-VIII-1862, *Miller* (K), IX-1879, *J. M. Hildebrandt 3125* (K), IV-1880, *3398* (K), *R. Baron 2225* (K), *2573* (K), *J. Blackburn* (K), 22-III-1923, *Radama 1617* (K, US).

MOZAMBIQUE. 18-IV-1959, *Barbosa & De Lemos 8537* (K), 22-II-1961, *A. Balsinhas 296* (K), 25-III-1961, *A. Gomes & Sousa 4668* (K), 20-II-1961, *Mogg 29744* (K), 16-X-1965, *32242* (K), *J. Stocks* (K).

SEYCHELLES. III, 1875, *J. Harne 476* (K).

TANGANYIKA (TANZANIA). Kisarwe District: 14-XII-1915, *A. Peters K254* (K); Lindi District, 6-IV-1935, *H. T. Schlieben 6240* (K), 26-II-1936, *B. D. Burtt, 5456* (K); Newalda District; Makonde Plateau, 19-VII-1941, *Gillman 1018* (K); Rufiji District; Magia Is., 23-III-1933, *G. B. Wallace 832* (K); Tanga District, VI-1893, *C. Holst 2901* (K); 12-X-1918, *Peters K409* (K), 5-XII-1935, *Burtt 5355* (A,K), 27-III-1957, *R.E.S. Tanner 3460* (K, NY), 3-II-1958, *3990* (K, NY); Kwale to Tanga, 31-I-1939, *P. J. Greenway 5840* (K); Dar-Morozoro, III-1965, *J. Procter 2948* (K), 20-VI-1914, *Peters K78* (K), 5-XII-1917, *K366* (K).

ZANZIBAR, ?ZANZIBAR IS., III-1861, *J. Kirk* (K), II-1874, *Hildebrandt 1217* (K).

"EAST AFRICA." 28-X-1928, *Greenway 933* (K), 18-VII-1930, *2295* (K), *C. F. Elliott* (K), 27-28-XII-1911, *E. Heller* (US), 13-III-1946, *Crower & Sousa 3370* (K), 11-III-1928, *T. Y. Wigg 398* (A,K,US).

Vernacular names: "msandarusi," "mtandarusi," "munyama" (Kenya).

In Madagascar and Mauritius the flowers of *H. verrucosa* may have five subequal, well-developed petals or only three, a variation that is probably genetic (Brenan, 1967). Brenan also noted that flowers in East African specimens of this species all have three well-developed and two minute petals, and appear to be tripetalous. This difference in number of well-developed petals was one of the main reasons in recognizing several *"Trachylobium"* species. We, however, agree with Oliver's (1871) and Gilg's (1897) view that they are all synonymous with *H. (Trachylobium) verrucosa.*

This species may be separated from its closest relatives in the New World, *H. oblongifolia* and *H. parvifolia,* by its (1) highly verrucose-rugose fruit, (2) partially glabrous ovary, and (3) occurrence of tripetalous flowers.

Trachylobium gaertnerianum was described from a single specimen collected by Martius in Java. Whether or not this specimen came from an indigenous tree has long been in dispute. Bentham and Hooker (1865) apparently considered it to be indigenous but thought it conspecific with one of the African Trachylobia. Oliver (1871) believed the specimen to have come from a cultivated tree. Gilg (1897), from referring to several native names listed by Hasskarl (1842) and Miquel (1855), believed this "species" indigenous in Java. Until further information is available, we assume that the natural range of *H. verrucosa* is restricted to Africa.

2. *Hymenaea oblongifolia* Huber, Bol. Mus. Paraense Hist. Nat. 5: 386, 1909 (Figures 8, 12, 13)

Large trees 20-40 m. tall, the trunk up to 100 cm. in diameter, the buttress roots *ca.* 1.5 m. tall and 1 m. wide, or lacking, the bark smooth, silvery grey with fine, longitudinal cracks. Leaves firmly petiolate, the petiole 2-2.5 cm. long, the stipules narrowly elliptic, *ca.* 20-40 x 4-8 mm., obtuse, glabrous, or densely tomentose outside and glabrous inside, caducous, the leaflets with twisted petiolule *ca.* 8 mm. long, the petiolule glabrous or pubescent, the lamina narrowly to broadly oblong, occasionally falcate, 12-16 x 4-5 cm. (or 2-4 times this size in saplings and young trees), coriaceous, glabrous and shining on both surfaces but duller beneath or sparsely tomentose above and densely golden brown tomentose beneath, obtuse to short-acuminate, the base oblique, with the inner side acute and the outer side slightly to moderately rounded, the midrib and secondary veins sharply prominent on abaxial surface, the veinlets obscure. Inflorescence densely paniculate, the racemes terminal and lateral, 6-12 cm. long, the branches yellowish brown to golden brown tomentose, slender, rather elongate (to 5 cm. long),

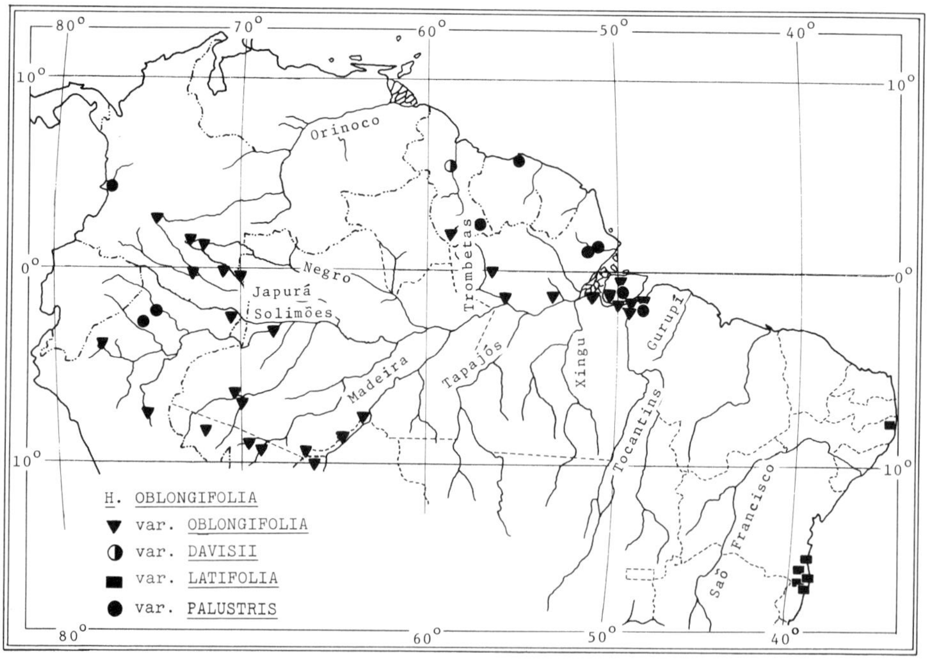

12. Distribution of *Hymenaea oblongifolia.*

slightly flexuous, the pedicels 2-3 mm. long, thickly articulate at apex, bracts and bracteoles suborbicular, *ca.* 5 x 4 mm., caducous. Flowers relatively small, the hypanthium disc-bearing, with a stalklike base 3-9 mm. long; calyx lobes narrowly to broadly ovate, 7-12 x 3.5-4.5 mm., slightly acute to obtuse, subequally tomentose on both sides; petals creamy white or rose-tinged, oblanceolate to broadly spatulate, subsessile to distinctly unguiculate, glabrous; stamens *ca.* 18-25 mm. long, the anthers elliptic, 2-4 mm. long; ovary shortly stipitate, oblique-obovoid, compressed, pubescent throughout, densely hirsute at base and extending as ridges toward apex, the stipe *ca.* 2-3 mm. long, the style attached laterally, 10-15 mm. long, glabrous, with a minute capitate stigma. Fruit ovoid to obovoid, subcompressed, 3.0-7.0 x 2.5-4.5 x 2.5-3.5 cm., or occasionally larger, the suture not prominent. Seeds 1 or 2, occasionally 3, avoid to ellipsoid, the testa brown to dark brown, its intrusion into the cotyledons producing 2 or 4 furrows.

Key to Varieties of *Hymenaea oblongifolia*

1. Leaflets glabrous on both sides.
   2. Leaflets narrowly falcate; petals lanceolate, subsessile; hypanthium with a stalklike base *ca.* 9 mm. long . . . . B. var. *davisii*
   2. Leaflets narrowly to broadly oblong; petals subunguiculate to distinctly unguiculate; hypanthium with a stalklike base, *ca.* 3 mm. long.
      3. Leaflets narrowly oblong (L/W *ca.* 3); petals subunguiculate to unguiculate, the claws to 2 mm. long . . . . A. var. *oblongifolia*
      3. Leaflets broadly oblong (L/W *ca.* 2); petals distinctly unguiculate, the claws to 8 mm. long . . . . C. var. *latifolia*
1. Leaflets densely golden brown tomentose beneath . . . . D. var. *palustris*

Hymenaea oblongifolia has the widest distribution of any species within the evergreen-forest ecosystems. Only recently has it been considered as a polymorphic species including four varieties (Lee and Langenheim, 1973). These varieties are separated morphologically by foliar characters, and isolated ecologically by distinctive habitats. Further details are given under the respective varieties.

Hymenaea oblongifolia may be separated from the African *H. verrucosa* by smooth pods and oblong leaflets; it is distinguished from *H. parvifolia* by glabrous petals, oblong leaflets, and more obovoid fruits.

2A. *Hymenaea oblongifolia* var. *oblongifolia* (Figures 12, 13)

Cynometra zamorana Schultes, Harvard Bot. Mus. Leafl. 13(10):301-302. Based upon *R. E. Schultes 5429,* Colombia: Comisaria del Vaupés, Rio Macays, vicinity of Cachivera del Diablo. 14-V-1943 (A, holotype!, IPEAN!, isotype).

Large trees to 30 m. tall, the trunk 65(+) cm. in diameter, the branchlets glabrous, with numerous lenticels. Leaves rigidly petiolate, the petioles 20-25 mm. long, glabrous, the leaflets with twisted petiolule *ca.* 6 mm. long, the lamina elongate-oblong to slightly falcate, 12-14 x 4-5 cm., coriaceous, glabrous, and shining, obtuse to short-acuminate, the base oblique, with the inner side acute and the outer side slightly rounded, the midrib and secondary veins prominent on abaxial surface, the veinlets obscure. Inflorescence long-paniculate, with terminal and lateral racemes 8-12 cm. long at maturity, its branches rather long and slender, slightly flexuous, brownish-puberulent, pedicels *ca.* 3 mm. long, the bracteoles *ca.* 5 x 3.5 mm. Hypanthium campanulate, with the stalklike base *ca.* 3 mm. long; calyx lobes broadly ovate, 7-8 x 4-6 mm., golden yellow sericeous-tomentose on both sides subequally; petals obovate to spatulate, *ca.* 12 x 5 mm., obtuse, unguiculate (claws *ca.* 2-3 mm. long) to subsessile; stamens *ca.* 18 mm. long, the anthers *ca.* 3 mm. long; ovary with style *ca.* 15 mm. long. Fruit obovoid to ovoid, slightly compressed 4.5-5.0 x 2.5-3.0 x 2.5 cm., smooth, the suture not evident. Seeds 1 or 2, ovoid to obovoid, *ca.* 25 x 15 x 14 mm., the testa dark brown, with 2 intrusions into the cotyledons producing 2 horizontal furrows. Chromosome number 2N = 24.

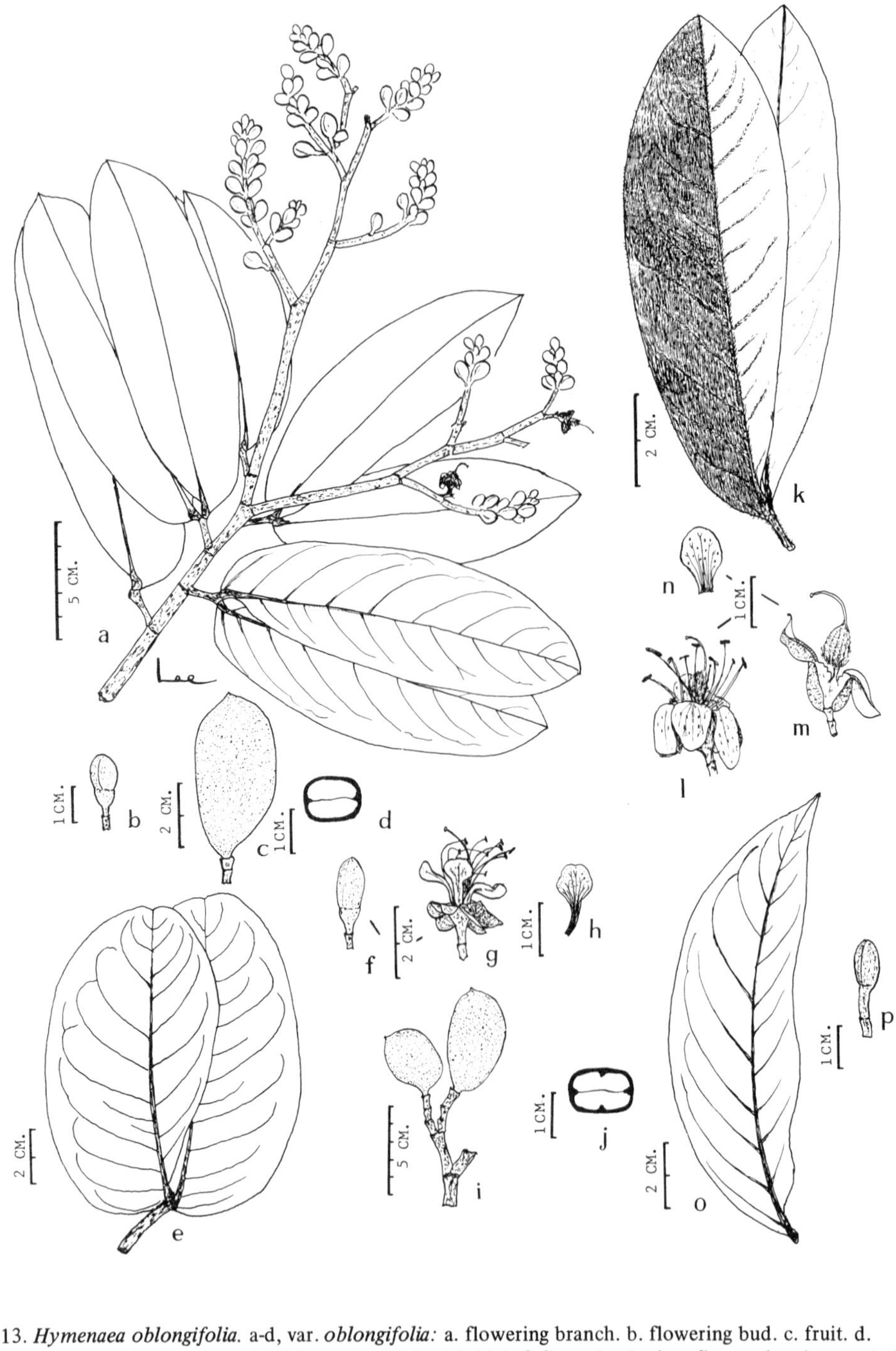

13. *Hymenaea oblongifolia.* a-d, var. *oblongifolia:* a. flowering branch. b. flowering bud. c. fruit. d. seed (cross-section). e-j, var. *latifolia:* e. leaf (adaxial side). f. flowering bud. g. flower showing erected petals. h. petal. i. fruits. j. seed (cross-section). k-n, var. *palustris:* k. leaf. l. flower. m. partly dissected flower showing position of stipe relative to hypanthium. n. petal. o-p, var. *davisii:* o. leaflet (abaxial side). p. flowering bud.

Type: "Habit in silvis ripariis fluminis Mapuera, infra locum Taboleirinho dictum. Ab. indigenis 'Jutahy' appellature." 13-XII-1907, *A. Ducke 9037* (MG, holotype!; RB!, isotype).

Habitat and distribution: This variety has the widest distribution of any *Hymenaea* species restricted to the Amazon and adjacent basins. Its distribution pattern conforms to high annual rainfall areas (more than 2500 mm.) of the eastern and western zones of the *hylaea* (Fig. 12). It is characteristic of the *várzea alta,* although several western collections were reported to be growing on *terra firme*. This confusion is understandable, because it is more difficult to draw a clear line between *terra firme* and *várzea* habitats in the western Amazon than it is on the eastern side. It flowers from September to October in the eastern zones of the Amazon and adjacent basins and about a month later in the western zones. Fruit falls about August.

Selected specimens: BRAZIL. Pará. Belém and vicinity: 15-IX-1948, *A. Ducke 2164* (A, COL, F, IPEAN, MG, P, R), 12-IX-1922, *16907* (K, RB, U, US), 3-VII-1947, *J. M. Pires 614* (COL, INPA, IPEAN, NY, US), 10-X-1952, *4332* (INPA, IPEAN, MG, NY), 17-VIII-1967, *Pires & Silva 10712* (IPEAN, MG), XII-1960, *Pires et al. 51871* (NY, UB): along Belém-Brasilia Highway, 2-V-1960, *E. Oliveira 663* (MG): Highway BR-17, 18-X-1955, *Ducke 21273* (MG); Boa Vista, Rio Tapajós, 24-VIII-1932, *Capucho 402* (F, IPEAN); Rio Cuminá, Lago Salgado, 15-X-1913, *Ducke 14982* (MG); Poco, 10-VIII-1966, *J. Elias 271;* Gurupá, 3-X-1916, *Ducke 16570* (MG, RB, US); Almerim, Rio Jari, 21-IV-1963, *Oliveira 2686;* Monte Dourado, 12-XI-1967, *3588* (MG), 27-IV-1968, *4331* (MG); Pirelli, VI-1958, *Pires 5956;* Marajó, Breves, 30-VII-1930, *Pires et al. 5834;* São Miguel dos Macacos-Breves, 20-XI-1971, *Lee 73* (UC).

Amazonas. Basin of Rio Juruá, near mouth of Rio Embira, 28-VI-1933, *B. A. Krukoff 5051* (A, LP, M, NY, U, UC, US); near Tres Casas, IX-X-1934, *6323* (A, F, NY, U, US); São Paulo de Olivença, Rio Solimões, 7-X-1913, *Ducke 24283* (K, RB, U, US).

Terr. Acre. Basin of Rio Juruá, VIII-1933, *Krukoff 5472* (A, F, K, M, NY, U, UC, US); Rio Muru, 6 Km. above confluence with Rio Taranaca, 16-IX-1968, *G. T. Prance et al. 7282* (A, INPA, MG, NY, P, R, U, US).

Terr. Rondônia. 70-100 km. along highway E of Porto Velho, 21-IX-1963, *B. Maguire et al. 56737* (F, NY).

BOLIVIA. Pando, Rio Abuna, 16-XI-1968, *Prance et al. 8519* (INPA, MG, NY, R, US).

COLOMBIA. Vaupés. Alto Vaupés, 6-I-1944, *Gutierrez & Schultes 524* (NY); Rio Apaporis, 1-15-XII-1951, *H. Garcia-Barriga 13847* (COL, NY, US); Macoys River, 14-V-1943, *Schultes 5424* (A, IPEAN, K, NY); Serra de la Macarena, Rio Guapaya, 30-XI-1949, *Phillison et al. 1624* (COL); Intendenacia de Meta, Serra de la Macarena, Rio Guapaya, 30-IX-1949, *1647* (IPEAN, US); Rio Caquita la Redrera, 30-XI-1912, *Ducke 12345* (MG, R, RB, US); Caña Guacaya, 24-IV-1952, *Schultes & Cabrera 16257* (A, U, US).

GUYANA. Basin of Essequibo River, near mouth of Onoro Creek, 15-24-XII-1937, *A. C. Smith 2813* (A, F, K, NY, U, US).

PERU. Loreto. Florida, Rio Putumayo at mouth of Rio Zubinata, X-XII-1931, *G. Klug 2291* (A, F, K, US), *J. Schunke 2129* (NY); Prov. Coronel Portillo, 13-VIII-1963, *R. L. Magin 71* (NY, US), *95* (F, NY, US); Rio Santiago, 29-X-1962, *J. J. Wurdack 2483* (A, F, K, NY, P, R, UC, US).

SURINAM. Baurent á Cayenne, 18-IX-1954, *L'Echantillon 214M* (P).

Vernacular names: "Jutaí," "jutaí de várzea," "parcunba" (Pará, Brazil); "loka" (Surinam); "yutubanco" (Loreto, Peru); "coca" (Vaupés, Colombia), this name is also widely used to refer to the completely unrelated narcotic plant, *Erythroxylon coca* Lam., in the same area.

This variety is characterized by its (1) narrowly oblong (L/W *ca.* 3) and glabrous leaflets, (2) the subunguiculate to subsessile petals, and (3) much softer wood.

2B. *Hymenaea oblongifolia* var. *davisii* (Sandwith) Lee & Lang., J. Arnold Arbor. 54(1): 102-3. 1973 (Figures 12, 13)

Hymenaea davisii Sandwith, Bull. Misc. Inf., Roy. Bot. Gard. Kew 1931: 366-367.

Tree *ca.* 35 m. tall, the trunk *ca.* 60 cm. in diameter. Leaves glabrous throughout, rigidly petiolate, the petiole *ca.* 25 mm. long, the leaflets with twisted petiolule *ca.* 5 mm. long, the lamina narrowly oblong and falcate, 11-15 x 3-4.2 cm., glabrous and shining, short- to long-acuminate, the base oblique, with the inner side acute and the outer side slightly rounded. Inflorescence paniculate, to 15 cm. long, its branches light brown puberulent, the pedicels 5-7 mm. Hypanthium campanulate, with a stalklike base *ca.* 9 mm. long; calyx lobes elliptic-oblong, pale brown tomentose outside, densely golden brown sericeous inside; petals oblanceolate to lanceolate, 13-15 x 5-6 mm., subsessile; stamens 20-30 mm. long, the anthers elliptic, 4-5 mm. long; ovary with style 20-30 mm. long, the ovules up to 6. Fruit ovoid, 4.0-4.5 cm. long, 3.5 cm. wide, usually one-seeded.

Type: In true Wallaba *(Eperua)* forest, along Warimia Creek, Essequibo River, Guyana, VI-1929, *Forestry Dept. Record No. 943* (K, holotype!).

Habitat and distribution: This variety has been collected only twice, both times along streams in the Wallaba forest, in the drainage of the Essequibo River. Flowering occurs in June.

Specimens examined: GUYANA. Macouria River, Essequibo River, III-1909 (with fruit), *Anderson 165* (K).

Vernacular names: "Locust," "simiri," or "courabarie," although the first two names refer also to the more common *H. courbaril* in Guyana.

Hymenaea davisii was named by Sandwith in honor of Mr. T. A. W. Davis, then Assistant Conservator of Forests, British Guiana. Sandwith recognized its close affinity to *H. oblongifolia* but separated *H. davisii* by its (1) narrowly falcate and long-acuminate leaflets, (2) long pedicels, (3) long, stalklike hypanthium base, and (4) usually one-seeded fruit. We concluded that the differences listed are to be considered merely as trends within a rather diverse species complex. Thus, we concluded that it was appropriate to reduce *H. davisii* to varietal status under *H. oblongifolia* (Lee and Langenheim, 1973).

2C. *Hymenaea oblongifolia* var. *latifolia* Lee & Lang. J. Arnold Arbor. 54(1): 98-101. 1973 (Figures 8b, 12, 13)

Evergreen to facultatively deciduous tree *ca.* 25 m. tall, the trunk *ca.* 1 m. in diameter. Leaves glabrous, the petioles rigid, 2-3 cm. long, the stipules narrowly elliptic, obtuse, 15-20 mm. long, 5-8 mm. wide, caducous, the leaflets with twisted petiolule *ca.* 6 mm. long, the lamina broadly oblong, 8-13 x 4.5-7 cm. (L/W *ca.* 2), glabrous and shining, obtuse to retuse or occasionally short-acuminate, the base oblique, with the inner side acute and the outer side moderately rounded, the midrib and secondary veins sharply prominent beneath. Inflorescence paniculate, the racemes *ca.* 18 cm. long, the branches golden brown puberulent, *ca.* 5 cm. long at base, the bracteoles *ca.* 5 x 3 mm. Hypanthium campanulate, with a stalklike base *ca.* 4 mm. long; calyx lobes narrowly ovate to oblong, slightly acute, 9-12 x 6 mm. long, 6 mm. wide, about equally tomentose on both sides; petals broadly spatulate, *ca.* 15 x 9 mm., the lamina 8-9 mm. in diameter, unguiculate (claws *ca.* 8 mm. long), erect during anthesis; stamens *ca.* 25 mm. long, the anthers *ca.* 4 mm. long; ovary with short stipe *ca.* 3 mm. long, compressed oblique-obovoid, the style *ca.* 15 mm. long; ovules 3 or 4. Fruit ovoid to subcompressed-ovoid, 4-7 x 2.5-4.5 x 2.5-3.5 cm., occasionally larger, the suture not prominent, the calyx occasionally remaining attached at base. Seeds 1 to 3, rarely more, obovoid to ellipsoid, *ca.* 30 x 18 x 12 mm., the testa dark brown, *ca.* 0.2-0.4 mm. thick, with 4 intrusions into the cotyledons producing 2 horizontal and 2 vertical, shallow furrows. Chromosome number 2N = 24.

Type: Brazil. Bahia, Itabuna: along road between main entrance and library (Reserve Botânica) on grounds of CEPEC, I-1971, *R. S. Pinheiro 1720* (CEPEC, holotype!), 2-II-1972, *Lee 162,* 29-XII-1972, *194* (UC). All three specimens were collected from the same tree, which is still living on the grounds of CEPEC.

Habitat and distribution: This variety is restricted to the Atlantic coastal evergreen forest which has been heavily cut for cultivation of cacao. It is most frequently found as isolated trees in cacao plantations *(matta de cacau)* or along rivers (Fig. 8b) in the forest. It occurs particularly on sandy soils. The flowering season is from January to late February. Fruit falls about December.

Specimens examined: BRAZIL. Pernambuco: Rio Gurajaú, vicinity of Recife, XII-1951, *Ducke & Andrade-Lima 70* (IPEAN, R), XII-1952, *Andrade-Lima 52-1197* (IPA). South Bahia: village of Rio Branco, near Una, X-1969, *Langenheim 5648, 5649* (UC). Along east coast from Una to Prado, 3-10-II-1972, *Lee 166, 167, 168, 169, 170, 171, 173, 175, 176, 179, 187* (UC).

Vernacular names: "Jatobá burundanga" and "jatobá farinheira" (east cost of southern Bahia).

The variety was described in 1973 after studying material collected by JHL in 1969. The close affinity with the Amazonian species *H. oblongifolia* was immediately recognized, although the final determination was not made until flowers and complete inflorescences were collected by R. S. Pinheiro and YTL in 1972.

This ephithet expresses the exceptional width of the leaflets. The name should not be confused with *H. latifolia* Hayne (Hayne, 1830), which has been appropriately transferred to the genus *Peltogyne* (Bentham, 1870). The first two specimens of this variety were collected in the vicinity of Recife by the late Professor A. Ducke and Professor D. Andrade-Lima in the early fifties and were subsequently erroneously identified by Ducke as *H. latifolia* Hayne.

This variety is distinguished by (1) the broadly-oblong leaflets (length-width ratio about 2, as compared to 3 in var. *oblongifolia*), (2) the distinctly clawed petals, and (3) its habitat in uplands of the Atlantic coastal evergreen forest.

2D. *Hymenaea oblongifolia* var. *palustris* (Ducke) Lee & Lang., J. Arnold Arbor. 54(1): 101-2. 1973 (Figures 8a, 12, 13)

Hymenaea palustris Ducke, Ann. Jard. Bot. Rio de Janeiro 1: 24. 1915.

Trees 20 to 30 (-40) tall, the trunk *ca.* 85 cm. in diameter, the young branches sometimes rusty-tomentose. Leaves rigidly petiolate, the petioles *ca.* 15 mm. long, thinly puberulent, the leaflets with twisted petiolule *ca.* 6 mm. long, the petiolule densely pubescent, the lamina narrowly oblong, 12-16 x 3-4.5 cm., sparsely pubescent above, densely golden brown tomentose beneath (the leaflets may be glabrous on both sides in seedlings and saplings), acute to obtuse, the base oblique, with the inner side acute and the outer side slightly rounded. Inflorescence paniculate, its branches subgolden brown puberulent, the pedicels *ca.* 2 mm. long. Flowers smaller than other varieties, hypanthium with a stalk-like base *ca.* 2 mm. long; calyx lobes *ca.* 7 x 5 mm., obtuse to slightly acute, about equally white-rusty sericeous on both sides; petals oblanceolate, *ca.* 12 x 14 mm., subsessile to subunguiculate; stamens to 15 mm. long, the anthers *ca.* 3 mm. long; ovary with style *ca.* 12 mm. long, the ovules 3 or 4. Fruit ovoid to obovoid, *ca.* 4.5-5.5 x 2.5-3.5 x 2.0-2.5 cm., smooth, the suture not evident. Seeds 1 to 2, rarely 3, obovoid to ellipsoid, *ca.* 30 x 18 x 12 mm., the testa dark brown, its intrusion into the cotyledons producing 2 horizontal, shallow furrows.

Type: "in silvis, paludosis ad rivulorum margines," Utinga, near Belém, Pará, Brazil, 6-X-1914. *A. Ducke 15496* (MG, holotype!; RB!, US!, isotype).

Habitat and distribution: This variety is restricted to swamp forest or *igapó*; it derives its name from this habitat. It shares a distribution essentially similar to that of var. *oblongifolia*, but occurs in the *igapó* rather than the *várzea*. The flowering season is from September to October in Brazil and about a month later in Peru. Fruits fall in August.

Specimens examined: BRAZIL. Pará. Gurupu, 2-X-1916, *Ducke* (MG-16567) (K, MG, P, R, RB, U, US); Rio Pacaja and Rio Muirapiranga, 15-X-1965, *Prance et al. 1660* (IPEAN, K, NY, US).

Terr. Amapá. Along Rio Araguari, near Serra do Navio, 1-X-1961, *Pires et al. 51391, 51393* (IPEAN, MG), 2-X-1961, *51438* (A, COL, F, IPEAN, MG, R, RB, SP, UB, UC, US, VEN), 15-XI-1971, *Lee 67, 68, 69, 70* (UC).

COLOMBIA. Rio Yurumangui, Dept. de Valle, 14-II-1944. *J. Cuatracasas k5836* (COL, F, US).

GUYANA. Upper New River, 11-XII-1935, *T. G. Myers 5868* (K); *R. H. Schomburgk* (K).

PERU. Loreto, Rio Itaya, 7-X-1929, *L. Williams 3453* (F, US); Quebrada Valentin-Rio Tahuaya, IX-X-1963, *A. A. Vargas 131* (NY, US).

Vernacular names: "Jutaí" and "jutaí de igapó" (Pará and Terr. Amapá).

Ducke distinguished *H. palustris* from *H. oblongifolia* by such small differences as (1) hairs on lower leaf surface, (2) narrower leaflets, (3) smaller flowers, and (4) much harder wood. These differences are sufficiently minor so that Sandwith (1938) questioned whether *H. palustris* should not be made a variety of *H. oblongifolia.* In examining specimens for this study, we have felt justified in reducing *H. palustris* to varietal status under *H. oblongifolia* (Lee and Langenheim, 1973).

3. *Hymenaea parvifolia* Huber, Bol. Mus. Paraense Hist. Nat. 5: 385-6, 1909 (Figures 14, 15)

Medium-sized or small trees or occasionally shrublike, *ca.* 20 m. tall, 60 cm. in diameter, the young branches rather slender, glabrous. Leaves petiolate, the petioles slender, 1.5-2 cm. long; the leaflets with twisted petiolule *ca.* 3 mm. long, the lamina usually small, broadly falcate, 5-8 x 2.5-3 cm., occasionally much larger (up to 8-10 x 4-5 cm.), coriaceous, glabrous and shining on both surfaces but duller beneath, short-acuminate (or with a relatively long tip in seedlings and saplings), the base oblique, with the inner side acute and the outer side slightly to moderately rounded, the midrib prominent on abaxial surface, secondary and other veins obscure on both sides. Inflorescence a long panicle, composed of compound racemes 4 cm. long or much longer, its branches ochraceous-puberulent, the pedicels 5-6 mm. long, articulate in middle, the bracts and bracteoles suborbicular, to 4 mm. long, caducous. Flowers small, hypanthium disc-bearing, obconic, 3 mm. long; calyx lobes ovate, *ca.* 6 mm. long, subequal, ochraceous-tomentose outside, white-sericeous inside; petals creamy white, narrowly oblanceolate, 8 x 2.5-3.0 mm., obtuse, glabrous below, pilose above with rigid, white hairs; stamens usually 10, occasionally 12, 13, 14 or more (in such cases flowers and all parts relatively larger), the filaments *ca.* 28 mm. long, creamy white, the anthers *ca.* 3 mm. long; ovary stipitate, oblique-ovoid, pubescent throughout, densely hirsute at base and with ridges of short hairs extending toward apex, the style elongate, *ca.* 25 mm. long, the stigma minutely capitate. Fruit small, ovoid to obovoid, 3.5-4.5 x 2.4-2.8 x 2.0-2.5 cm., brown to dark brown, smooth, the suture ± evident, usually 1-seeded. Seeds ovoid, 25 x 16 x 12 mm., the testa brown to dark brown, with its intrusion into the cotyledons producing 2 or 4 rather shallow furrows.

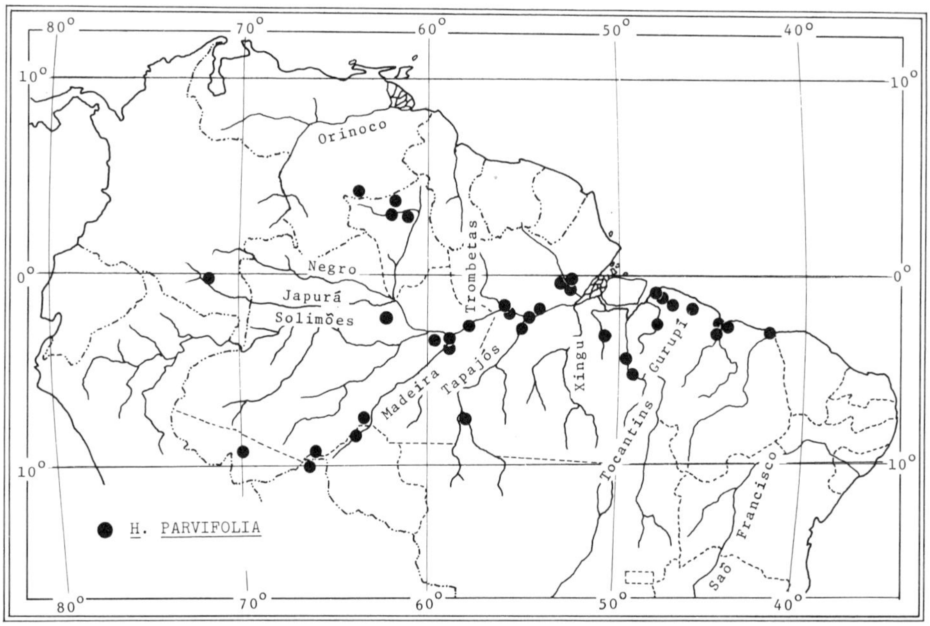

14. Distribution of *H. parvifolia.*

Type: Brazil. Belém, Pará: "insilvis paludosis ad rivulorum margines," 6-X-1915, *A. Ducke (MG-15496)* (MG, holotype!).

Habitat and distribution: Hymenaea parvifolia occurs primarily in the drier eastern and central areas with an annual rainfall of less than 2500 mm. In the eastern zone it occurs in coastal areas of Piauí and Maranhão and in forests of northeastern Pará. It has frequently been collected in *campo* near Monte Alegre and Almeirim, and occurs in *terra firme* near Santarém and Manaus, along Rio Madeira, in the northern part of Roraima, and has been collected once in Venezuelan Amazonia. It flowers from August to early October. Fruit falls about August.

Selected specimens: BRAZIL. Pará. Rio Itacainas, Serra Buritirana, 3-VIII-1970, *Pires & Belém 12490* (IPEAN), *12998* (IPEAN); Highway BR-22, Capanema to Maranhão, 2-XI-1965, *Prance & Pennington 1914* (F, IPEAN, NY, P, U); Alcântera, 10-IV-1954, *Fróes 30737* (US); Monte Alegre, 8-XI-1948, *Pires 1373* (IPEAN, NY), 11-XII-1908, *Ducke 9871* (IPEAN, P, RB, US), 4-IV-1909, *10185* (MG), 3-XII-1953, *Fróes & Filho 23950* (INPA), 5-V-1953, *Andrade-Lima 53-1351* (IPA, IPEAN). Santarém and its vicinity, *12 -IX-1969, Silva & Sousa 2563* (K, MG, NY, U, US), 23-X-1954, *Fróes 31226* (IPEAN, P, SP, US), 8-X-1969, *Langenheim & Rodrigues 5613, 5623, 5626, 5627, 5628* (INPA, UC), 2-XII-1966, *Cavalcante & Silva 1526* (IPEAN, MG); Belém and its vicinity, 28-IX-1959, *Kuhlmann & Jimbo 308* (IPEAN, SP, UB) 31-VII-1960, *Oliveira 958* (IPEAN, SP, UB), 22-X-1914, *Ducke 15511* (MG), 24-VII-1947, *Black 47-931* (INPA, IPEAN, NY, U, US); Obidos, 22-IX-1910, *Ducke 11041* (MG), *15111* (MG, R, RB, US); Mosqueiro, 19-X-1940, *Ducke 613* (F, IPEAN, MG, R, UC, US); Bragança, 6-XII-1908, *Ducke 9821* (MG, RB, US); Beltera, V-1954, *C. Townsend* (IPEAN); Cacaual Grande, 30-X-1950, *Black & Ledoux 50-10426* (COL, INPA, IPEAN); Breu Branco, Rio Tocantins, 27-IX-1949, *Fróes 23544* (A, COL, IPEAN, NY); Rio Piriá, Municipio de Vizeu, V-1958, *Fróes 31314* (IPEAN); Upper Cupary River, IX-1931, *Krukoff 1212* (NY); Region of Rio Jari, 5-XII-1967, *Oliveira 3846* (IPEAN, NY), 27-IV-1968, *4331* (IPEAN, NY).

Amazonas. Manaus and its vicinity, 2-X-1965, *W. Rodrigues 7629,* (INPA), 25-VI-1969, *Langenheim & Rodrigues 5618, 5619* (INPA, UC); Basin of Rio Madeira, Municipality Hymayta, near Tres Casas, 14-IX-X-1934, *Krukoff 6520* (A, US).

Maranhão. São Luiz and its vicinity, 6-XI-1948, *Ducke 2182* (IPEAN, R, US), II-III-1939, *Fróes 11613* (A, F, K, NY, U, UC, US), I-1940, *11711* (A, F, NY, U, US); Maracassumé River, *Fróes 1900* (A, F, K, LP, NY, P, U, US), 16-IX-1932, *1921* (A, P).

Mato Grosso. Near Tabajara, upper Machado River region, XI-XII-1931, *Krukoff 1379* (UC).

Piauí. Teresina, 26-VI-1972, *D. Sucre 9352* (RB, UC).

Terr. Roraima. Canto Galo, 21-I-1967, *Prance et al. 3951* (A, COL, F, MG, NY, R, U); Boa Vista, 10-III-1948, *Fróes 23056* (IPEAN, NY).

Terr. Acre. Near mouth of Rio Macauhau, basin of Rio Purus, 15-VIII-1933, *Krukoff 5519* (A, F, K, NY, U, UC).

BOLIVIA. Pendo. W. bank of Rio Madeira opposite Abuna, 10-VII-1968, *Prance et al. 5806* (A, INPA, NY, US).

COLOMBIA. Vaupés. Rio Apapôris, *Schultes & Cabrera 13003* (NY, US).

VENEZUELA. Bolivar, Rio Paramichi, 8-9-I-1962, *Steyermark 90768* (NY, US, VEN).

Vernacular names: "Jutaí mirim" (Amazonas), "jutaí pequeno," "jutaí pororoca," "jutobasinhó," and "jutahysinho" (Pará).

This is the only other member of sect. *Trachylobium* in the New World. It can be separated from *H. oblongifolia* by (1) the usually smaller, broadly falcate, short-acuminate leaflets, (2) the densely pilose hairs on the adaxial side of the petals, and (3) the usually smaller, ovoid, and one-seeded fruit. It can be distinguished from *H. verrucosa* by its smooth fruit, and pubescence of petals and ovary.

This species is commonly an emergent tree on sandy soil in *terra firme* forest, but it occurs also as small individuals in secondary forest (*capoeira*) or in grassy fields (*campo*). Ducke and Black (1953) noted that it is characteristic of several *campos* and that it might appear shrublike in the center of the *campo,* as a small tree at its periphery, or as a medium-sized to large tree in the forest. The fact that it has been found in rain forest, as well as transitional zones and drier fields, and that it exhibits different growth forms in various

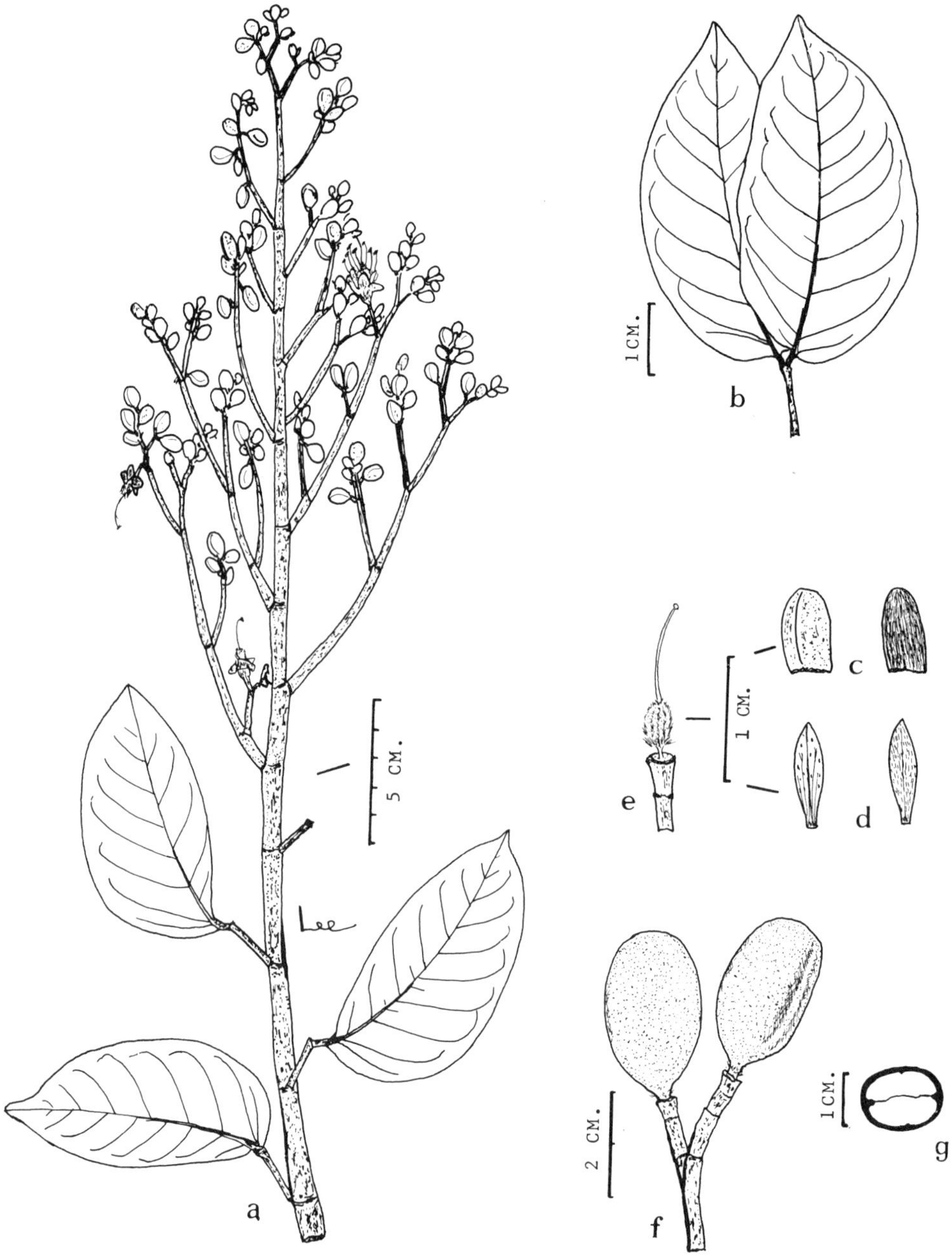

15. *Hymenaea parvifolia.* a. flowering branch. b. leaf (abaxial side). c. sepals. d. petals. e. pistil, hypanthium, and pedicel. f. fruits. g. seed (cross-section).

habitats may indicate great phenotypic plasticity and an adaptive trend toward a drier environment.

Of the specimens cited, those from Rio Jari and Santarém are worth mentioning. These are characterized by unusually large leaves, the size of leaflets occasionally reaching 8-10 x 4-5 cm., and the flowers are relatively larger and usually have more than 10 stamens. Despite these variations, they show no major differences from the "normal" population. We therefore regard them as not warranting special taxonomic status.

4. *Hymenaea torrei* León, Revista Soc. Cub. Bot. 6: 4. 1949 (Figures 16, 17)

Medium-sized tree *ca.* 10 m. tall, the branchlets light brown, flexuous, densely puberulent. Leaves short-petiolate, the petioles 7-10 mm. long, densely puberulent, the leaflets with twisted petiolule *ca.* 2 mm. long, the lamina elliptic to obovate-oblong, 2.5-4.7 x 1.5-2.5 cm., coriaceous, glabrous, and shining on both surfaces, obtuse to retuse, or occasionally short-acuminate, the base oblique, with the inner side acute and the outer side moderately rounded, the midrib and secondary veins prominent beneath, other veins finely reticulate and the margins straight. Inflorescence a short terminal panicle, densely corymbose, and with flexuous branches, the pedicels 4-5 mm. long, bracts and bracteoles caducous. Flowers relatively small, hypanthium campanulate, subsessile; calyx lobes suborbicular to oblong, 8-10 x 7-8 mm.; petals creamy white, obovate, 9-10 x 5-6 mm., sessile, hairy at base (?); stamens with red filaments, 16-20 mm. long, the anthers *ca.* 4 mm. long; ovary short-stipitate, obliquely and broadly oblong, densely pubescent, the style *ca.* 15 mm. long, the ovules up to 6. Fruit oblong to obovoid, rarely ovoid; 2.5-6.6 cm. long, 2.5-3.6 cm. wide, compressed, coarsely verrucose-rugose with numerous reddish lenticels on surface, slightly transversely ridged (showing seeds underneath). Seeds 1 to 4, the testa dark brown, 0.2-0.4 mm. thick, with its intrusion into the cotyledons producing 2 horizontal, rather shallow furrows.

Type: Cuba. Peninsula Cabo Cruz, Prov. Oriente, mouth of Rio Puercas, West of Pilon, VII-1949, *Brothers Alain & Crisógono 1051* (LS, holotype).

Habitat and distribution: This species is endemic to Cuba, restricted to Prov. Oriente and neighboring Camagüey where it has been reported growing at a river mouth, on "virgin hills," or on scarcely accessible cliffs.

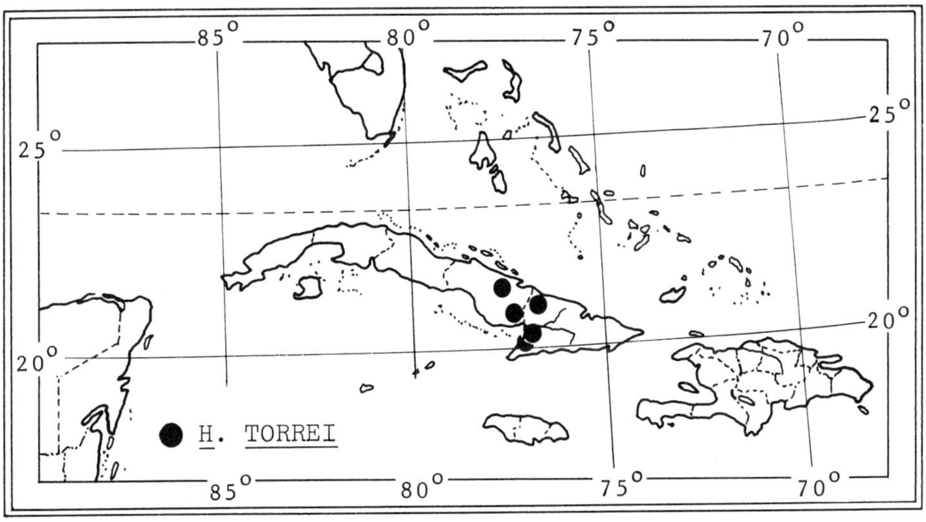

16. Distribution of *H. torrei.*

Specimens examined: CUBA. Prov. Oriente: Peninsula Cabo Cruz, Cape Cruz, Vedado, Habana, coastal thickets, on limestone dog-tooth rocks, VII-1935, *Brother León 16350* (A). South of Niquero, in limestone terraces facing sea, 16-I-1923, *E. L. Ekman 16165* (NY). Prov. Camagüey: North of Prov. Oriente, between Gloria and Piloto, 27-II-1909, *J. A. Shafer 583* (A, NY, US).

Due to the scarcity of available specimens we are listing below three other collections cited by León (1949) which we have not examined, but which are presumably in the Herbarium of College de La Salle (LS), Habana, Cuba.

Prov. Oriente. Near Puerto Padre, *Curbelo 273.* Prov. Camagüey: On barren hill near Cespedes, *Roig 897;* Caobilla, *Acuna 4033.*

Vernacular names: It is called "Caguaran amarillo" in Puerto Padre, Oriente, for the reddish yellow color of the wood. It is called "Caguraian" in Camagüey, and "Guasima italiana" in Pilon, Oriente, because of its similarity of fruit of the "Guasima" (*Copaifera* sp. ?).

Hymenaea torrei was named by Brother León in honor of the late Prof. Carlos de la Torre, a renowned naturalist and botanist of Cuba.

Although endemic to Cuba, the verrucose fruit of this species strongly suggests a tie to the presumed ancestral stock in Africa. In a detailed explanation of the present highly disjunct distribution of *Hymenaea* (Langenheim, 1973; Langenheim and Lee, 1974) two possible routes for oceanic dispersal are mentioned. *Hymenaea torrei* might be explained either as a result of the transportation of fruit from an African ancestral stock via the South Equatorial Current to the West Indies, or "torrei-type" populations might have first arrived on the shores of northern South America and then have been dispersed to the islands of the West Indies. In either case, however, the evolutionary history of *H. torrei* was probably independent of that of the Amazonian *H. oblongifolia* complex (Langenheim and Lee, 1974).

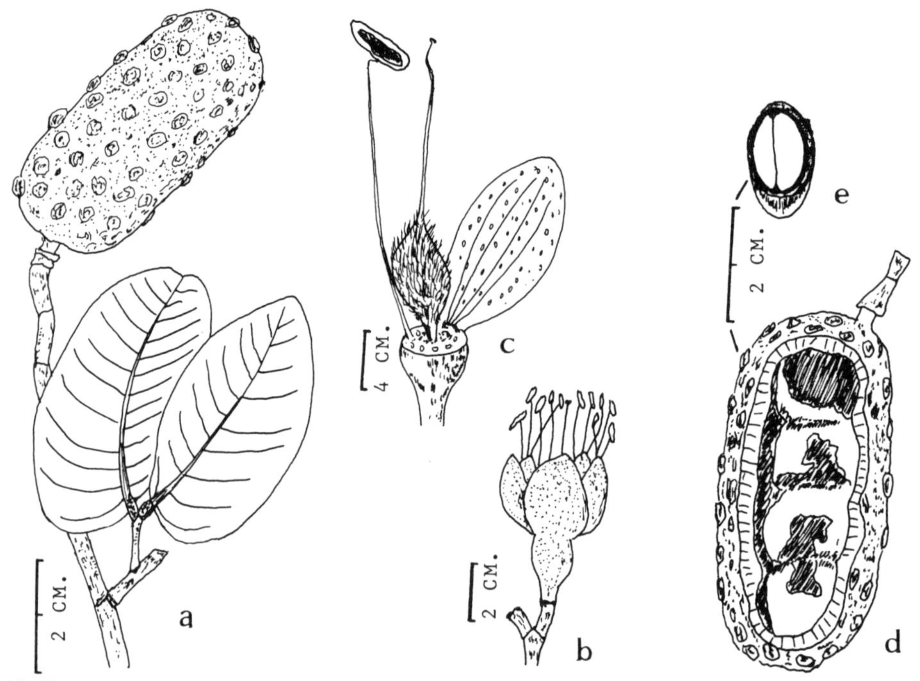

17. *Hymenaea torrei.* a. branch with fruit. b. flower. c. petal, stamen, and pistil. d. fruit, part of the pericarp removed showing endocarp and seeds. e. seed (cross-section). Redrawn from Léon (1949). (1949).

Despite apparent affinities with a presumed ancestral stock, *H. torrei* is placed at the beginning of sect. *Hymenaea* because of its type of inflorescence and certain floral characters, but it is not directly related to any other species in this section.

Hymenaea torrei may be distinguished from species in sect. *Trachylobium* by its rather short, corymbose inflorescences and from sect. *Hymenaea* by its verrucose fruit.

5. *Hymenaea eriogyne* Benth. in Mart. Fl. Bras. 15(2): 237. 1870 (Figures 8f, 18, 19)

Small shrubby and many-branched, deciduous trees *ca.* 5 m. tall, the young branchlets glabrate to finely puberulent. Leaves short-petiolate, the petioles 2-3 mm. long, the leaflets with twisted petiolule *ca.* 3 mm. long, the petiolule pubescent or not, the lamina ovate to obovate, 2-4 x 2.5-3 cm., occasionally larger, coriaceous, shining, usually subappressed tomentose above and densely pubescent beneath, less commonly glabrate on both surfaces, obtuse to retuse, the base oblique, with the inner side acute

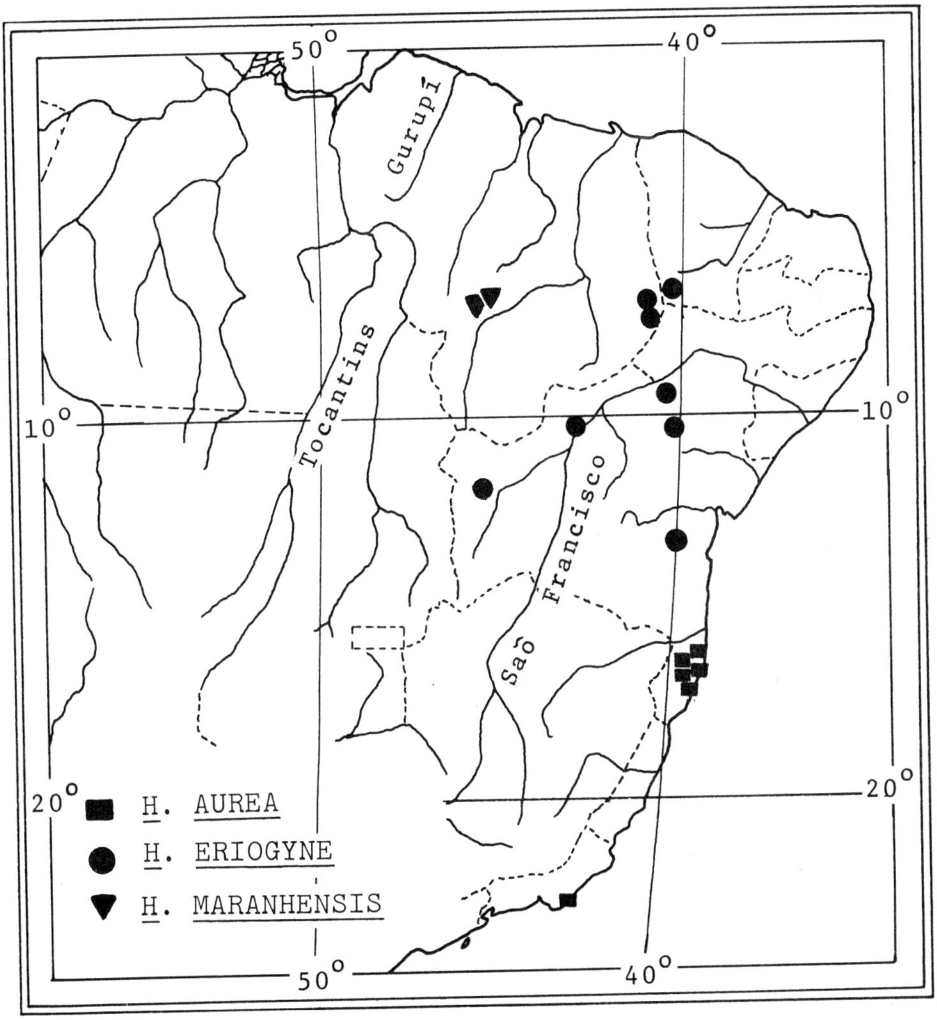

18. Distribution of *H. eriogyne*, *H. aurea*, and *H. maranhensis.*

and the outer side moderately rounded. Inflorescence composed of few racemes, its branches reddish brown puberulent, the pedicels 8-10 mm. long. Flowers medium-sized, hypanthium campanulate, with a stalklike base 4 mm. long; calyx lobes obovate, subequal, 15-25 x 12-17 mm., reddish to dark brown, velutinous-tomentose outside, golden brown sericeous within; petals spatulate, *ca.* 22 x 12 mm., distinctly clawed (claws to 10 mm. long); stamens 28 mm. long, the anthers *ca.* 8 mm. long; ovary long-stipitate, pilose with long, golden yellow hairs, moderately densely hirsute at base, the style 25-28 mm. long. Fruit light brown, oblong to subcylindrical, 9-14 x 3.5-4.5 x 3-3.5 cm., subcompressed, dotted with numerous lenticels, the suture not prominent. Seeds 4-8, ellipsoid, *ca.* 18 x 15 x 8 mm., the testa light brown, its intrusion into the cotyledons producing 2 shallow, horizontal furrows. Chromosome number 2N = 24.

Type: Villa de Barra, Bahia, Brazil. *Blanchet 3096* (K; lectotype!).

Three collections bearing the type number *Blanchet 3096* have been examined. On the basis of annotations on the sheets and on minor morphological differences, we have concluded that they represent trees from different localities in the same general area. We designate the Villa de Barra collection as the lectotype because it was collected from the type locality cited by Bentham (1870).

Habitat and distribution: This species is restricted to thickets in thorn forests *(caatinga)* of northeastern Brazil, which covers the four neighboring states: Bahia, Ceará, Pernambuco, and Piauí.

Specimens examined: BRAZIL. Bahia: Japira, *Blanchet 3096* (A, NY); Jacobina, *Blanchet 3096* (K); Calaceiras da Cortessa, 7-XI-1912, *Zehntner 529* (RB); vicinity of Juazeiro, 15-V-1918, *H. M. Curran 250* (A, US); Cortesia, *Zehntner 1930* (M); Valley of Rio das Ondas, 4-IV-1971, *H. S. Irwin et al. 31478* (NY, UC); Espigao Mestre, 2-III-1972, *W. R. Anderson et al. 36426* (NY, UC).

Ceará: 17-X-1934, *Luetzelburg 26320* (M).

Pernambuco: Serra do Araripe, Estrada Araripina-Crato, 25-VI-1952, *Andrade-Lima & Magalhães 52-1094* (IPA, UC).

Piauí: 10 km SE of Picos, 24-XI-1971, *Lee & Andrade-Lima 116* (UC); vicinity of Itainopolis, *121* (UC).

Vernacular names: "Jatobá de veado" and "jatobá."

The corymbose inflorescence, composed of a few racemes, and the floral and fruit characters of *H. eriogyne* secure its position in *Hymenaea.* The pubescence of the ovary strongly suggests a link to the more primitive type of ovary found exclusively in sect. *Trachylobium.* Although *H. eriogyne* has been recorded growing sympatrically with *H. velutina* in the *caatinga* of northeastern Piauí (Fig. 8f), it is considered to be a relictual species of the once more expanded Amazonian *hylaea.*

Hymenaea eriogyne, the coastal *H. aurea,* and the *cerrado* species *H. maranhensis* undoubtedly form a coherent group within sect. *Hymenaea.* They are characterized by a pubescent ovary, spatulate petals, and short primary internodes in the seedling (the last unknown in *H. maranhensis*). They can be distinguished from one another by (1) the vesture of the ovary, (2) fruit characters, and (3) the nature of the bracteoles.

Hymenaea eriogyne can be separated from the remaining species of sect. *Hymenaea* by its densely pubescent ovary and rather small and ovate to obovate leaflets.

6. *Hymenaea aurea* Lee & Lang. J. Arnold Arbor. 54(1): 96-98. 1973 (Figure 18)

Large, evergreen to facultatively deciduous tree to 28 m. tall, the trunk to *ca.* 90 cm. in diameter, the young branches, leaf buds, petioles, and leaves covered with golden yellow tomentum, golden brown upon aging, the older branches brown to silvery grey, with peeling epidermis. Leaves short-petiolate, the petioles 3-7 mm. long, the stipules narrowly elliptic, 15-20 x 5-8 mm., caducous, the leaflets with twisted petiolule *ca.* 3 mm. long, the petiolule pubescent, the lamina oblong-obovate, 5-9 x 2-4 cm., coriaceous, sparsely puberulent and shining above, densely pubescent beneath with golden brown tomentum, obtuse to apiculate, the base oblique, with the inner side acute and the outer side slightly rounded, the midrib and secondary veins prominent beneath and densely hirsute-tomentose. Inflores-

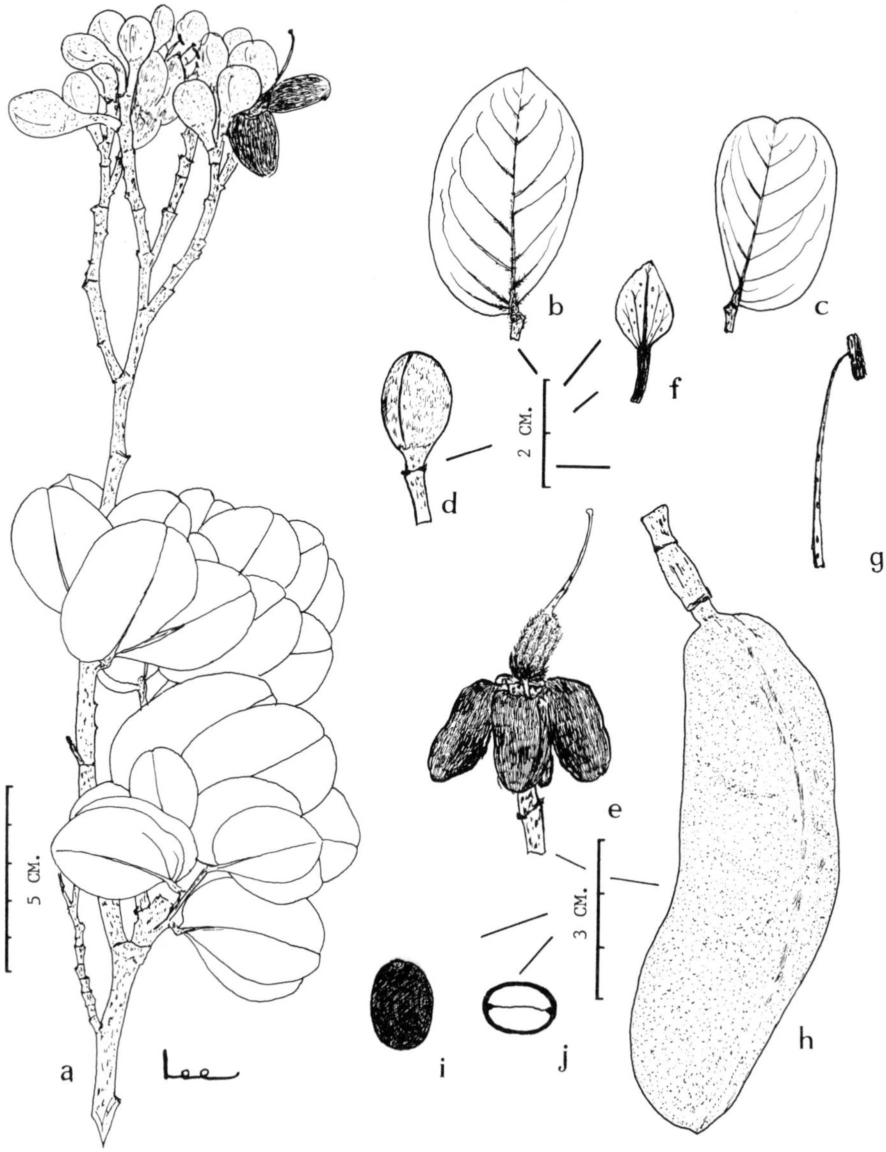

19. *Hymenaea eriogyne.* a. flowering branch. b, and c. leaflets. d. flowering bud. e. flower (with petals and stamens shed). f. petal. g. stamen. h. fruit. i. seed. j. seed (cross-section).

cence short-paniculate, corymbose during anthesis, its branches golden brown tomentose, the pedicel *ca.* 7 mm. long, densely tomentose; bracts and bracteoles caducous, densely golden brown puberulent dorsally, reddish brown and glabrous within. Flowers medium-sized, hypanthium campanulate, with a stalklike base *ca.* 9 mm. long; calyx lobes obovate, *ca.* 15 x 9 mm., golden brown tomentose dorsally, golden yellow sericeous inside; petals creamy white, spatulate, *ca.* 21 x 6 mm., distinctly unguiculate (claws *ca.* 9 mm. long), caducous; stamens *ca.* 30 mm. long, the anthers *ca.* 6 mm. long; ovary long-stipitate, oblong-oblique, heavily pubescent with golden yellow, woolly tomentum, the stipe 3-6 mm. long, the style *ca.* 25 mm. long, ovules 9-12. Fruit large and flat, 9-15 x 7-10 x 2.5-3.5 cm., light brown, conversely ridged and grooved (showing seeds underneath), the suture not evident. Seeds

5-8 (+), obovoid-ellipsoid, compressed, 32-35 x 20 x 12-15 mm., the testa dark brown, its intrusion into the cotyledons producing 2 shallow, horizontal furrows. Chromosome number 2N = 24.

Type: Brazil. Bahia: Una, *ca.* 3 km. east of Fazenda São Raphael, 28-X-1969, *Jean H. Langenheim 5644* (UC, holotype!).

Habitat and distribution: This species has been reported to be restricted to the Atlantic coastal evergreen forest along the east coast of southern Bahia (Lee and Langenheim, 1973), where it was found in the same area as *H. oblongifolia* var. *latifolia.* The two rarely occur together; however, one older collection *(RB 788)* from the state of Rio de Janeiro has been identified as *H. aurea,* so the species has a wider distribution than had been thought. The flowering season is from late December to early February. Fruits fall from November to December.

Specimens examined: BRAZIL. Bahia: southern Bahia: Fazenda São Raphael, Una, 28-X-1969, *Langenheim 5645, 5646, 5647* (UC); east coast from Una to Eunápolis, 2-12-II-1972, *Lee 163, 164, 177, 178, 180, 181, 183, 184, 186* (UC). Rio de Janeiro. Restinga do Camorim? Jacarepagúa, *RB-788* (RB).

Vernacular names: "Jatobá piloso" and "jatobá verdadeiro" (east coast of southern Bahia).

Hymenaea aurea is characterized by the dense, golden yellow tomentum on young branches, leaf buds, petioles, leaves, and flowers, and the specific epithet is derived from this pubescence. This species can be distinguished from *H. eriogyne* and *H. maranhensis* by (1) the ovary, which is densely tomentose but never hirsute at the base, (2) by its unusually flat fruit, and (3) by its growth habit as a large tree in the coastal evergreen forest.

7. *Hymenaea maranhensis* Lee & Lang. J. Arnold Arbor. 55(3): 441-2. 1974 (Figure 18)

Small tree, usually shrubby, and branched near base, 2-5 m. tall. Leaves petiolate, the petiole 12-16 mm. long, densely pubescent, the leaflets with twisted petiolule *ca.* 5 mm. long, coriaceous, dull and sparsely covered with appressed tomentum above, densely pubescent beneath with reddish brown, matted tomentum (the hairs most prominent on veins and margins), obtuse to retuse, the inner side acute and the outer side rounded, the midrib and secondary veins evident on both sides, elevated beneath and densely tomentose. Inflorescence composed of few racemes, each relatively few-flowered, its branches subgolden brown pubescent, the pedicels 8-14 mm. long, bracts not seen, the bracteoles ovate to obovate, 16-18 mm. long, 14-16 mm. wide, usually persistent during anthesis. Flowers relatively large, the older bud oblong, *ca.* 28 x 14 mm., hypanthium campanulate, subsessile; calyx lobes oblong, 25-30 x 10-15 mm., freyish brown tomentose outside, tomentose-sericeous within; petals spatulate, 28 x 12 mm. distinctly unguiculate (claws 6-7 mm. long), early deciduous; stamens 30-35 mm. long, the anthers *ca.* 8 mm. long; ovary long-stipitate (the stipe *ca.* 8 mm. long), obliquely oblong, *ca.* 10 x 4 mm., densely yellowish pilose at base and along the two rims but usually bare in central parts, the style *ca.* 25 mm. long, glabrous, the ovules 8-12. Mature fruit and seeds not seen.

Type: Brazil. Maranhão: "Muncipio de Loreta, 'Ilha de Balsas' region, between Balsas and Parnaibá rivers, *ca.* 20 km. south of Loreto, north of main house of Fazenda Trabalhosa," April 8, 1962, *G. and L. T. Eiten 4183* (NY, holotype!; UB!, isotype, additional type material was sent to K, SP and US from UB).

Habitat and distribution: This species is restricted to southeastern Maranhão, and it has been collected only three times. It grows on sandy soil in xeromorphic woodland, a type of *cerrado* (regionally it is called "*chapada*"). The flowering season is from February to April. Mature fruit has not been collected.

Specimens examined: BRAZIL. Maranhão, Municipio de Loreto: "Ilha de Balsas" region between Balsas and Parnaibá rivers, 38 miles south of Loreto, Feb. 3, 1970, *G. and L. T. Eiten 10460* (NY, UB) *ca.* 1 km. north of Parnaibá River at Santa Barbara, Feb. 18, 1970, *10669* (UB).

Vernacular names: "Jatobá do vaqueiro" and "jatobá da quaresma" (Maranhão, Brazil).

Professors G. Eiten and L. T. Eiten have kindly sent us their *Hymenaea* collections from Maranhão for study, from which was drawn the description of this species (Lee and Langenheim, 1974a). The specific epithet is derived from the state of Maranhão.

This species is closely related to *H. eriogyne* and *H. aurea,* but can be separated from them by (1) the unique vesture of the ovary (densely pilose at base and along margins but glabrate in center), (2) the two relatively large and persistent bracteoles, (3) the long pedicels, and (4) the sessile hypanthium. It can be distinguished from the other species of sect. *Hymenaea* by its pubescent ovary.

Hymenaea maranhensis may be easily confused with the more common *cerrado* species, *H. stigonocarpa* (particularly to its variety *pubescens*), by its shrubby habit and thickly coriaceous leaves. The two species, however, can be separated by the vesture pattern of the ovary and the shape of petals.

8. *Hymenaea reticulata* Ducke, Bull. Mus. Hist. (Paris) 4(6): 726. 1932 (Figures 8c, 20, 21)

Large evergreen to faculatatively deciduous trees *ca.* 25 m. tall, the trunk *ca.* 35 cm. in diameter, the young branchlets glabrous. Leaves glabrous throughout, rigidly petiolate, the petiole 2-3 cm. long, the leaflets with twisted petiolule *ca.* 8 mm. long, the lamina slightly falcate to oblong, 11-18 x 5-7.5 cm., semicoriaceous, glabrous but scarcely shining on both surfaces, short-acuminate, the base oblique, with the inner side acute and the outer side slightly rounded, the midrib, secondary veins, and densely reticulate veinlets prominent on both surfaces. Inflorescence short-paniculate, corymbose during anthesis, its branches glabrous, the pedicels 8-12 mm. long, bracts and bracteoles suborbicular, *ca.* 9 x 7 mm., caducous. Flowers medium-sized, hypanthium campanulate, with a stalklike base *ca.* 8 mm. long; calyx lobes obovate, 10-14 x 5-7 mm., subequal, ochraceous tomentose outside, greyish white sericeous inside; petals narrowly oblanceolate, *ca.* 12-14 x 8-11 mm., subequal, obtuse, short-clawed, stamens *ca.* 20 mm. long, the anthers elliptic, 6 mm. long; ovary short-stipitate, glabrate or with a few hairs at base, the stipe 3-4 mm. long, the style 16-18 mm. long, the ovules 6-12. Fruit oblong-rhomboidal, strongly compressed, large, 10-12 x 6-10 x 2.5-3.5 cm., the surface smooth, feltlike, with-

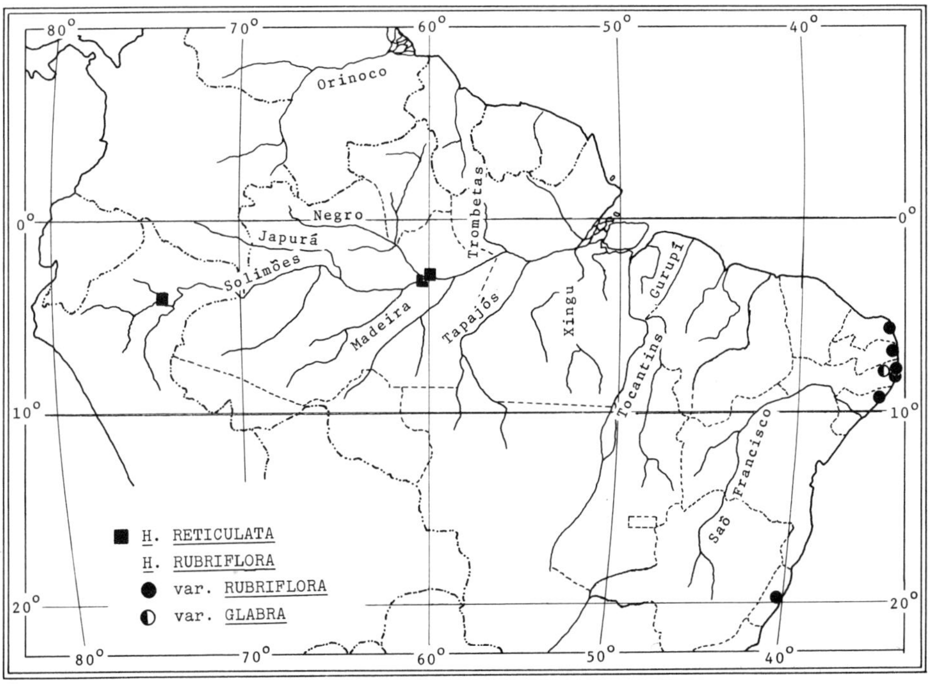

20. Distribution of *H. reticulata* and *H. rubriflora.*

out apparent resinous pockets or lenticels, reddish brown, occasionally with transverse ridges and grooves showing seeds beneath, the suture not evident. Seeds 3-8(+), ellipsoid, *ca.* 35 x 20 x 15 mm., the testa dark brown, its intrusion into the cotyledons producing 2 horizontal and 2 vertical, shallow furrows. Chromosome number 2N = 24.

Type: Brazil. Amazonas, Manaus: "loco Estrada do Aleixo, silva non inundabili ad ripas paludosa rivuli." Nov. 2, 1929, *A. Ducke (RB-23281)* (RB, holotype!; P!, U!, US!, isotypes).

Habitat and distribution: Hymenaea reticulata was thought until recently to be restricted to a small area in the vicinity of Manaus, where a few large trees occur in the *terra firme* forest on sandy soil or along stream margins. A collection, previously determined as *H. oblongifolia* from near Iquitos, Loreto, Peru, appears to be *H. reticulata* on the basis of both floral and foliar characters. Thus the species occupies a wider range than previously supposed. The flowering season is from September to October. Fruit falls in August to September.

Specimens examined: BRAZIL. Amazonas, Manaus, and vicinity: 17-I-1941, *Ducke 666* (F, IPEAN, MG, NY, R, UC, US), 2-IX-1965, *Rodrigues 7629* (INPA), 9-IX-1966, *Prance et al. 3075* (F, A, INPA, MG, NY, R, US, U), 8-VIII-1968, *Aluisio 88* (INPA), 14-VIII-1968, *Rodrigues et al. 8517* (INPA, UC), 2-X-1969, *Langenheim & Rodrigues 5617* (INPA, UC), 22-X-1971, *Lee 49, 51* (UC).

PERU. Dept. Loreto, Iquitos, V-1930, *G. Klug 1339* (F, NY, US).

Vernacular name: "Jutaí," a name also applied to other *Hymenaea* species.

Hymenaea reticulata has previously been considered to be closely related to *H. courbaril.* The oblanceolate petals and type of seedling suggest that it is more closely linked to the coastal *H. rubriflora* than to *H. courbaril.* It may, however, be separated from *H. rubriflora* by its relatively large, slightly acuminate, glabrous, and prominently reticulate-veined leaflets and its large, oblong-rhomboidal fruit with a smooth, feltlike surface.

9. *Hymenaea rubriflora* Ducke, Mem. Inst. Oswaldo Cruz 51: 457. 1953 (Figures 20, 22)

Small to medium-sized evergreen to facultatively deciduous trees 5-15 m. tall. Leaves short- to long-petiolate, the petioles 6-12 mm. long, densely pubescent or glabrous, the leaflets with twisted petiolules 3-5 mm. long, the petiolule pubescent or not, the lamina oblong, 7-15 x 3.5-7.5 cm., thick or semicoriaceous, sparsely pubescent and shining above, densely golden brown tomentose beneath, or glabrous on both surfaces, obtuse or apiculate, the base oblique, with the inner side acute and the outer side slightly to moderately rounded, the midrib and secondary veins prominent, densely tomentose beneath or not. Inflorescence a short panicle, composed of few racemes, each relatively few-flowered, its branches densely reddish brown puberulent, the pedicel 4-8 mm. long, bracts and bracteoles suborbicular, *ca.* 10 x 9 mm., caducous. Flowers medium-sized, hypanthium broadly campanulate, subsessile; calyx lobes ovate-obovate, obtuse, 12-18 x 8-12 mm., subequally pubescent on both surfaces with fine, golden brown tomentum; petals red (very rarely creamy white), narrowly lanceolate, obtuse, 14-24 x 5-8 mm., short-unguiculate (claws *ca.* 2 mm. long); stamens with filaments mostly red, *ca.* 25-38 mm. long, the anthers 4-6 mm. long; ovary short-stipitate, the stipe 2-5 mm. long, obliquely oblong, compressed, glabrous throughout, the style *ca.* 30 mm. long, the ovules 9-12.

Key to Varieties of *Hymenaea rubriflora*

1. Leaflets semicoriaceous, glabrous on both sides . . . . A. var. *glabra*
1. Leaflets thick-coriaceous, densely golden brown tomentose beneath . . . . B. var. *rubriflora*

The diagnostic characters suggest a close relationship of this species to the Amazonian *H. reticulata. Hymenaea rubriflora* differs from the latter by its (1) thickly coriaceous, densely pubescent or occasionally glabrous, and less coriaceous leaflets, (2) smaller and rugose fruit, and (3) usually red petals and filaments.

The most striking feature of this species, as implied by its specific epithet, is the red color of petals and filaments, but this does not necessarily hold in all populations. One collection from Paraíba *(Andrade-Lima 67-4988)* which from other characters clearly belongs

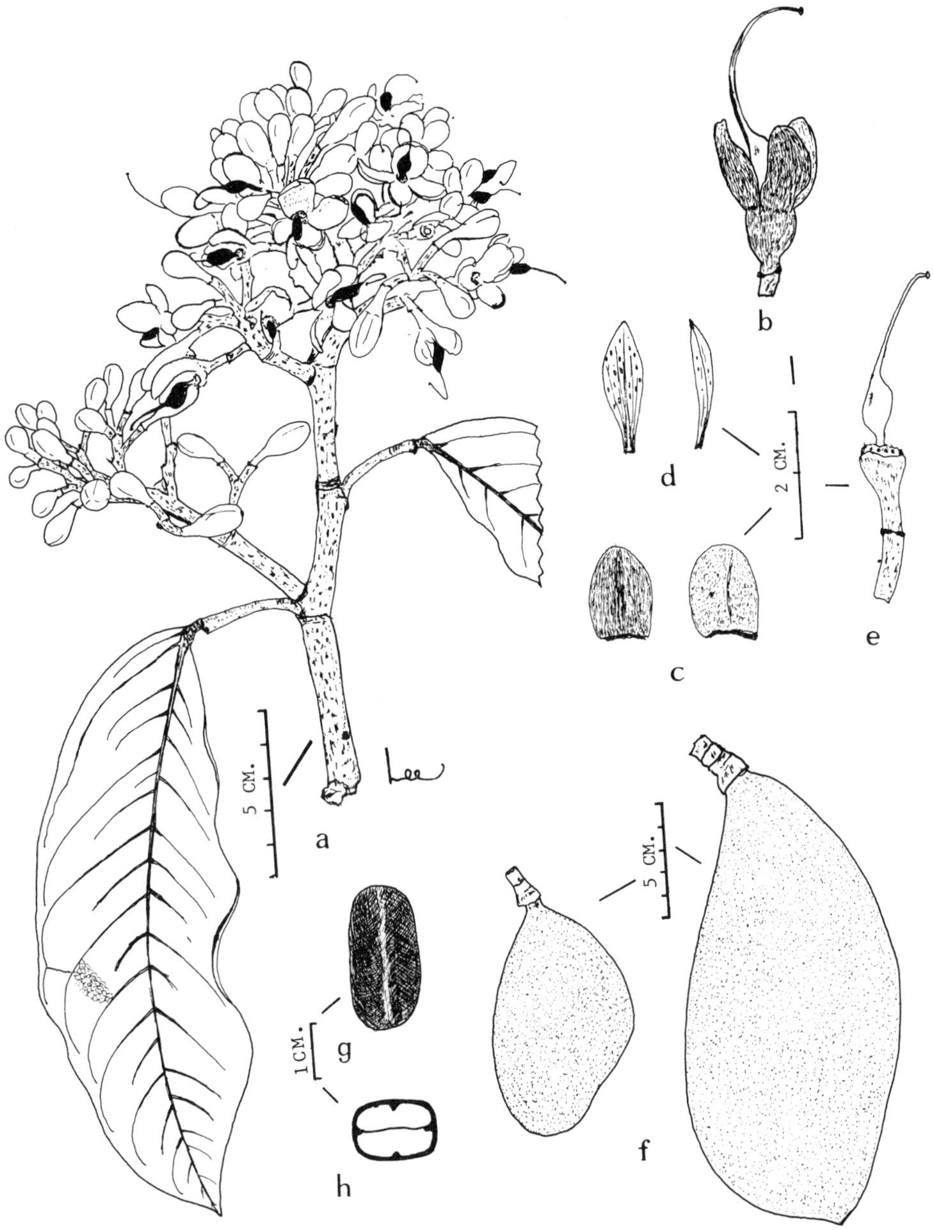

21. *Hymenaea reticulata.* a. flowering branch. b. flower (with petals and stamens shed). c. sepals. d. petals. e. pistil, hypanthium, and pedicel. f. fruits. g. seed. h. seed (cross-section).

to this species, shows creamy white petals and filaments. Although this difference in color might be due to aging, it is more likely that color varies in different populations.

Hymenaea rubriflora may be distinguished from other species in this section by (1) usually red petals and filaments, (2) large, densely pubescent or occasionally glabrous leaflets, (3) lanceolate petals, and (4) oblong, rather narrow, compressed fruit.

9A. *Hymenaea rubriflora* var. *rubriflora* (Figures 20, 22)

Small to medium-sized trees 5-15 m. tall. Leaves short-petiolate, the petioles 6-9 mm. long, densely pubescent, the stipules 30-45 x 8-12 mm., densely tomentose outside, glabrous inside, early-deciduous, the leaflets with twisted petiolule *ca.* 3 mm. long, the lamina obliquely oblong, narrower than in var. *glabra,* 7-10 x 3.5-5 cm., thick-coriaceous, sparsely pubescent and shining above, densely golden brown tomentose beneath, the midrib and secondary vein prominent on both surfaces, elevated beneath, the margins of mature leaves inrolled. Pedicels *ca.* 5-8 mm. long. Hypanthium subsessile; calyx lobes 12-18 x 10-12 mm.; petals red (very rarely creamy white), narrowly lanceolate, obtuse, 20-24 x 6-8 mm., briefly unguiculate (claws *ca.* 2 mm. long); stamens with filaments mostly red, *ca.* 28 mm. long, the anthers 5-6 mm. long; ovary short-stipitate, obliquely oblong, the stipe 2-4 mm. long, the style *ca.* 30 mm. long, the ovules 9-12(+). Fruit oblong, 5-6 x 3 x 1.5-2 cm., moderately compressed, light to dark brown, the suture not prominent. Seeds 3-6(+), oblong to ellipsoid, *ca.* 20 x 15 x 7 mm., the testa dark brown, its intrusion into the cotyledons producing 2 shallow, horizontal furrows. Chromosome number 2N = 24.

Type: Brazil. Pernambuco, vicinity of Recife, "in silvula secundária (*capoeira*) terra altis ad Estrada de Aldeia," 30-IV-1952, *Ducke & Andrade-Lima 99* (IPA, holotype! ; IPEAN!, SP!, isotypes).

Habitat and distribution: This variety has been known only from evergreen forests near Recife. Recent collections extend its range northward to Rio Grande do Norte and southward to Espirito Santo and Rio de Janeiro. Thus it has the widest distribution of any *Hymenaea* restricted to coastal forest.

Specimens examined: BRAZIL. Pernambuco, Recife, and vicinity: 27-VI-1950, *C. G. Leal* (RB), 6-VI-1949, *Andrade-Lima 49-267* (IPA, IPEAN, RB), 19-VIII-1949, *49-280* (IPA), 15-V-1955, *55-2056* (IPA), 17-X-1969, *Langenheim 5632* (UC), 14-XII-1971, *Lee 127* (UC). Aliança, Matta do Cravo: 4-II-1971, *P. Ferreira 56-71* (IPA).

Paraíba. João Pessoa, Alto do Cabo Branco: 16-IV-1967, *Andrade-Lima 67-4988* (IPA).

Rio Grande do Norte. Baia Formosa: 5-VII-1959, *L. Emygdio 1678* (R, UC).

Espirito Santo. Victoria: 20-III-1934, *RB 34205* (RB). Mata de Tabuleiro do Quirino, Reserva de Sooretana: 22-VII-1969, *D. Sucre 5952* (RB, UC).

Alagoas. Maceio: 1929, *Tupinambá 1238* (RB).

Vernacular name: "Jatobá," a name also applied to other *Hymenaea* species in Pernambuco.

9B. *Hymenaea rubriflora* var. *glabra* Lee & Andrade-Lima, J. Arnold Arbor. 55(3): 444. 1974 (Figures 20, 22)

Small tree, 5-6 m. tall. Leaves petiolate, the petioles slender, 12-20 mm. long, glabrous, the leaflets with twisted petiolule *ca.* 4 mm. long, the lamina broadly oblong, 10-14 x 4-7 cm., semicoriaceous, glabrous, and shining, obtuse to apiculate, the base oblique, with the inner side acute and the outer side moderately rounded, the midrib and secondary veins prominent, elevated beneath, the veinlets finely reticulate and conspicuous on both surfaces, the pedicels *ca.* 4-6 mm. long, bracteoles *ca.* 10 x 8 mm. Hypanthium obconical, sessile to subsessile; calyx lobes oblong, obtuse, *ca.* 16-18 x 6-9 mm.; petals red, lanceolate, *ca.* 14-15 mm.; stamens *ca.* 25 mm. long, the anthers *ca.* 4-5 mm. long; ovary long-stipitate, obliquely oblong *ca.* 7-8 x 3.5 mm., the stipe *ca.* 5 mm. long, the style *ca.* 20 mm. long, the ovules 6-10(+). Fruit and seeds not seen.

Type: Brazil. Pernambuco: "Bonito, Mata secund., marg. estr. 1 km. alem divisa Camocim de São Felix," 10-II-1967, *D. Andrade-Lima 67-4929* (IPA, holotype!, UC!, isotype).

Habitat and distribution: This new variety is known solely from the type collection which was made in secondary forest in hills along the Atlantic coast. The flowering time of the type collection was February.

Vernacular name: "Jatobá."

This new variety may be easily separated from var. *rubriflora* by the glabrous, much broader, and less coriaceous leaflets.

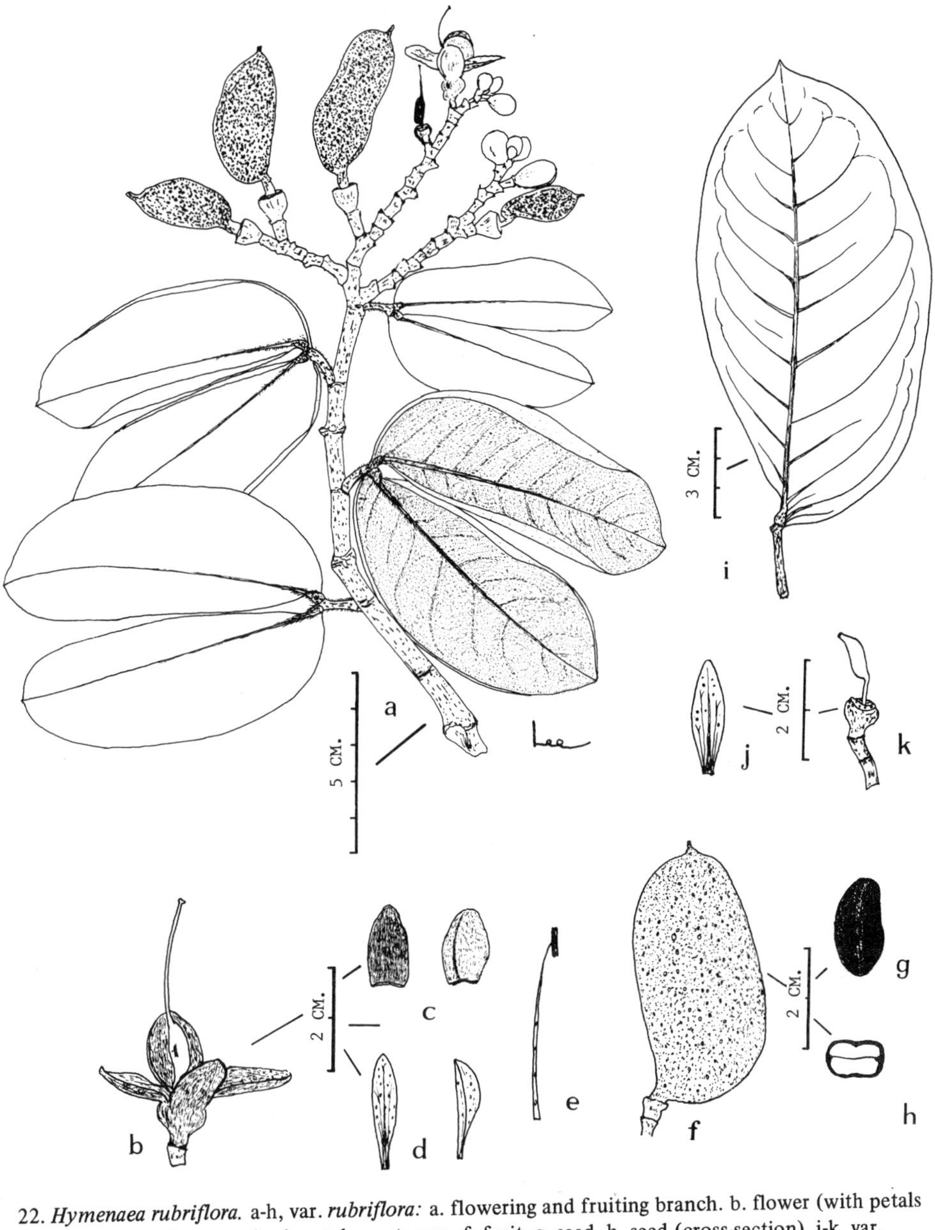

22. *Hymenaea rubriflora.* a-h, var. *rubriflora:* a. flowering and fruiting branch. b. flower (with petals and stamens shed). c. sepals. d. petals. e. stamen. f. fruit. g. seed. h. seed (cross-section). i-k, var. *glabra:* i. leaflet (abaxial side). j. petal. k. pistil, hypanthium, and pedicel.

The oblong, rather large (*ca.* 14 x 7 cm.), and glabrous leaflets are similar to those of *H. oblongifolia* var. *latifolia,* which has been collected along Rio Gurjau, Pernambuco, and in southern coastal Bahia. The two species, however, belong to different sections of the genus and can be separated by the sectional differences.

10. *Hymenaea intermedia* Ducke, Arch. Jard. Bot. Rio de Janeiro, 3: 92. 1922 (Figures 23, 24)

Large trees, 20-45 m. tall. Leaves rigidly petiolate, the petioles *ca.* 4-8 mm. long, glabrous to densely pubescent, the leaflets with twisted petiolule *ca.* 4 mm. long, the petiolule glabrous to lightly pubescent, the lamina oblong to obovate, occasionally subfalcate, 6-12 x 2.5-4.5 cm., coriaceous, glabrate, and shining on both surfaces, or densely pubescent beneath, acute to short-acuminate, the base oblique, with the inner side acute and the outer side moderately to strongly rounded, the midrib and secondary veins prominent and elevated beneath, glabrate to densely pubescent, the veinlets obscure. Inflorescence a short panicle, densely corymbose during anthesis, many-flowered, its branches ochraceous to rusty puberulent, the pedicels *ca.* 3-8 mm. long, bracts and bracteoles early-deciduous. Flowers small (older buds *ca.* 9-13 x 6-8 mm.), hypanthium campanulate, with a stalklike base *ca.* 3-8 mm. long, disc-bearing, additional nectariferous tissue not evident; calyx lobes obovate to oblong, ochraceous to rusty pubescent outside, rusty to golden brown tomentose-sericeous inside; petals creamy white, oblanceolate to obovate, *ca.* 9-12 x 3-4.5 mm., subunguiculate to sessile; stamens *ca.* 25 mm. long, the anthers 2-3 mm. long; ovary borne on a slender stipe, 2-4 mm. long, characteristically rhomboidal, strongly compressed, glabrate or occasionally with a few pilose hairs on one side at base, the style attached laterally, usually on the side opposite the stipe, filiform, *ca.* 25 mm. long. Mature fruit rhomboidal. Seeds 1 or 2, rarely more.

Key to Varieties of *Hymenaea intermedia*

1. Leaflets oblong, glabrate or occasionally sparsely puberulent beneath; base straight, not rounded-protracted . . . . A. var. *intermedia*
1. Leaflets obovate to oblong, lightly to densely pubescent beneath; base rounded-protracted on one side . . . . B. var. *adenotricha*

Hymenaea intermedia is now restricted to the Amazon Basin. One leaflet and possibly some sepals preserved in Oligo-Miocene amber found in Chiapas, Mexico have been studied by Miranda (Langenheim, 1966). The "characteristic nature of the leaflet" indicated *Hymenaea,* and Miranda noted that form and size of leaflet were somewhat similar to *H. in-*

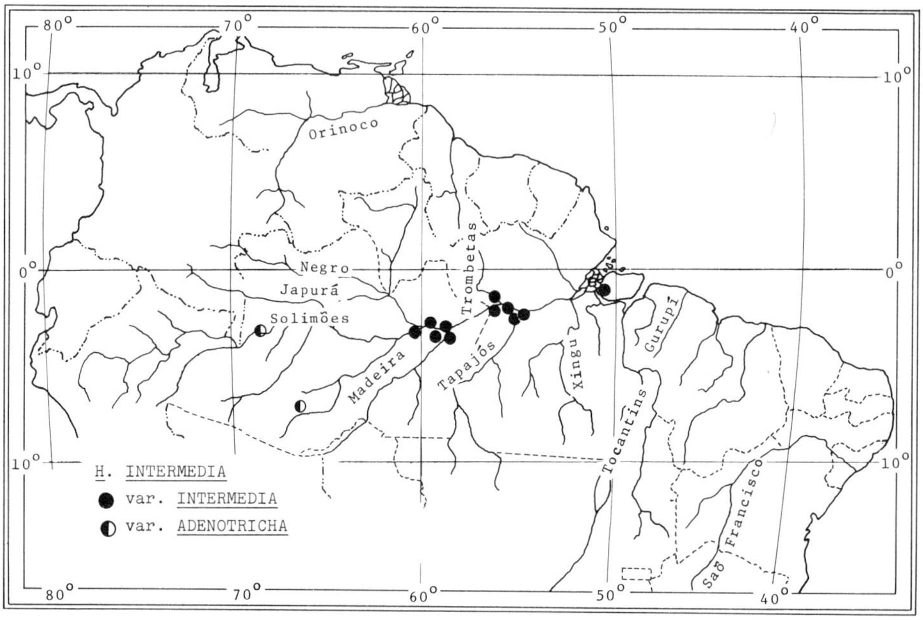

23. Distribution of *H. intermedia.*

termedia but that the glands (resinous pockets) were the same as in specimens of *H. courbaril.* The "sepals" appeared to him to belong to *Hymenaea* because of the pubescence and glands. These "sepals" differ from sepals of *H. courbaril* in lacking pubescence on the inside. It is worth pointing out that bracts and bracteoles of all living *Hymenaea* are glabrous inside and densely pubescent outside. The sepals, on the other hand, are characterized by dense pubescence on both sides. These inclusions in amber thus appear to be bracts or bracteoles. Their early-deciduous nature would enhance their chances of being "trapped" in the resin. Unfortunately, the specimens have not been available for further study.

10A. *Hymenaea intermedia* var. *intermedia* (Figures 23, 24)

Large trees 25-45 m. tall, *ca.* 55 cm. in diameter, the bark smooth, silvery grey. Leaves with petioles *ca.* 8 mm. long, glabrate to lightly puberulent, the leaflets with twisted petiolule 2-4 mm. long, the lamina oblong to obovate, occasionally subfalcate, 6-12 x 2.5-4.5 cm., glabrate and shining on both surfaces, but occasionally sparsely pubescent near the base and along the margins, the base oblique, with the inner side acute and the outer side slightly rounded, the midrib and secondary veins prominent beneath, but obscure above. Inflorescence a short panicle, densely corymbose during anthesis, its branches ochraceous-puberulent, the pedicel 7-8 mm. long, bracts and bracteoles *ca.* 5 x 4 mm. Flowers larger than in var. *adenotricha* (older buds *ca.* 13 x 9 mm.), hypanthium with a stalk-like base *ca.* 5-8 mm. long; calyx lobes 9-12 x 6-8 mm., with short ochraceous tomentum outside, densely white to yellow tomentose-sericeous inside; petals *ca.* 9-12 x 3.0-4.5 mm., subunguiculate; stamens *ca.* 25 mm. long, the anthers *ca.* 2-3 mm. long; ovary stipitate, rather small (*ca.* 2 x 2 mm.), and distinctly rhomboidal, glabrate throughout or occasionally with a few hairs at base or along edge, the stipe 2-4 mm. long, the style filiform, *ca.* 22-25 mm. long, the ovules 2-6. Mature fruit rhomboidal, occasionally obovoid to ovoid, 4-6 x 3-4 x 3.0 cm., the stalk short and acentric. Seeds 1 or 2, obovoid, the testa reddish to dark brown, its intrusion into the cotyledons producing 2 shallow, horizontal furrows. Chromosome number 2N = 24?

Type: Brazil. Pará: "Habitat inter Obidos et Trombetas ad rivulum silvestrem montis Curuma radicibus proximum," 1-X-1915, Ducke (MG-15778) (MG, holotype!; RB!, US!, isotypes).

Habitat and distribution: Variety *intermedia* is an emergent tree in primary *terra firme* forests in the eastern and central Amazon Basin, from Ilha do Marajó to Manaus. The flowering season is from September to October. Fruit falls about August.

Specimens examined: BRAZIL. Pará: Bella Vista, Rio Tapajós, 12-IX-1916, *Ducke 16487* (MG, P, R, RB); Porto Velho, 31-V-1952, *G. A. Black et al. 52-14645* (IPEAN); Vila Braga, Rio Tapajós, 13-IX-1922, *Ducke 16903* (RB), 23-IX-1922, *16908* (RB, U); Breves, Ilha do Marajó, 30-VII-1956, *Pires et al. 5520, 5854* (IPEAN).

Amazonas. Manaus and vicinity: Ducke Reserve, 28-VIII-1968, *INPA-21402* (INPA, UC), 24-X-1968, *Aluisio 225* (INPA), 2-X-1969, *Langenheim & Rodrigues 5614, 5615* (INPA, UC), 14-IX-1965, *Rodrigues & Loureiro 7167* (INPA, UC), 13-VII-1966, *Rodrigues & Osmarino 8180* (INPA), *Rodrigues & Coelho 8339* (INPA), 18-II-1949, *Lanjow & Lindeman 2191* (NY), 18-22-X-1961, *Lee 45, 55, 58, 61* (UC); Manaus, 9-IX-1940, *Ducke 623* (F, IPEAN, MG, NY, R, UC, US), 17-XI-1929, *23280* (RB, US), 8-VIII-1955, *INPA 1594* (HB, INPA, MG), 17-X-1929, *Killip & Smith 30150* (A); Manaus-Itacoatiara, 24-V-1966, *Rodrigues & Coelho 7862* (INPA); Rio Taruma, 25-VII-1949, *T. Guedes 60* (A, MG, IPEAN); Rio Urubu, 16-IX-1949, *R. L. Fróes 25237* (IPEAN, SP, US).

Vernacular names: "Jutaí" and "jutaí mirim" (Amazonas and Pará).

Hymenaea intermedia var. *intermedia* is distinguished by its leaf and fruit characters which are intemediate between those of *H. courbaril* and *H. oblongifolia.* This might suggest that it represents an intermediate form from which *H. courbaril* could be derived.

This variety differs from *H. oblongifolia* in its glabrate ovary and corymbose inflorescence, and from *H. courbaril* in its much smaller flowers, narrower leaflets, and relatively

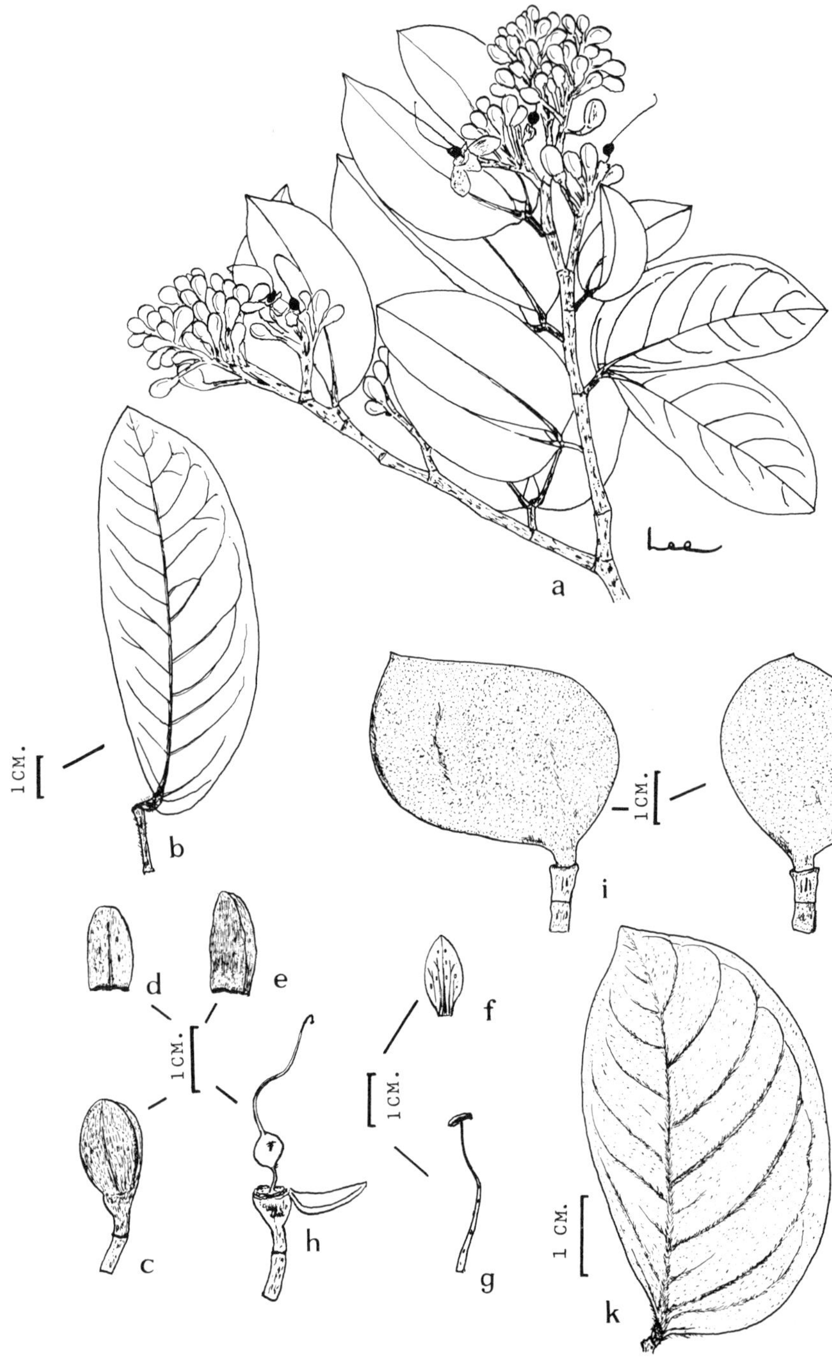

24. *Hymenaea intermedia.* a-j, var. *intermedia:* a. flowering branch. b. leaflet (abaxial side). c. flowering bud. d. sepal (abaxial side). e. sepal (abaxial side). f. petal. g. stamen. h. ovary, hypanthium, and pedicel. i. two-seeded fruit. j. one-seeded fruit. k-l, var. *adenotricha:* k. leaflet (abaxial side). l. ovary, hypanthium, and pedicel.

small, few-seeded fruit, which is rhomboidal when fully developed. It occupies drier habitats characteristic of *H. courbaril* rather than the moister one typical of *H. oblongifolia.*

10B. *Hymenaea intermedia* var. *adenotricha* (Ducke) Lee & Lang. J. Arnold. Arbor. 55(3): 450. 1974 (Figures 23, 24)

Hymenaea adenotricha Ducke, Bull. Mus. Hist. Nat. (Paris), ser. 2, 5: 727. 1932.

Large tree. Leaves rigidly coriaceous, short petiolate, the petioles *ca.* 6 mm. long, rusty short-to-mentose, the leaflets with twisted petiolule *ca.* 3 mm. long, the petiolule densely puberulent, the lamina obovate to oblong, slightly falcate, 5-7 x 2.5-4 cm., sparsely puberulent and shining above, densely pubescent beneath with rusty to subgolden tomentum, obtuse to apiculate, occasionally short-acuminate, the base oblique, with the inner side acute and the outer side broadly expanded (but less extreme than in *H. velutina*), the midrib and secondary veins prominent beneath, elevated, lightly to densely tomentose. Inflorescence rather short, *ca.* 4.5 cm. long, relatively few-flowered, its branches rusty-puberulent, the pedicels *ca.* 3-4 mm. long. Flowers smaller than in var. *intermedia* (older flower buds *ca.* 9 x 6 mm.), hypanthium with a stalklike base *ca.* 3 mm. long; calyx lobes *ca.* 9 x 7 mm., rusty golden brown tomentose subequally on both surfaces; petals *ca.* 9 x 3 mm.; stamens *ca.* 25 mm. long, the anthers *ca.* 2-3 mm. long; ovary stipitate, small (2 x 2 mm.), rhomboidal, strongly compressed, glabrate but often stiffly short hairy on one side at base, the stipe slender, *ca.* 3 mm. long, the style filiform, *ca.* 20 mm. long. Fruit (according to Ducke) similar to that of var. *intermedia.* Chromosome number unknown.

Type: Brazil. Amazonas: "Habitat silva non inundabili prope São Paulo de Olivença," 19-VIII-1929, *Ducke RB 23282* (RB, holotype!; K!, P!, isotypes).

Habitat and distribution: This variety has been known solely from the type collection made in São Paulo de Olivença, Brazil, close to the Peruvian border, where only a single specimen has been collected. A recent collection appears to belong to this variety, indicating a wider range than previously recorded. The flowering time of the type collection was August. Fruit and seeds have not been collected.

Specimens examined: BRAZIL. Amazonas: Rio Curuquete, vicinity of Cachoeira, 15-VII-1971, *Prance et al. 14342* (NY).

Hymenaea adenotricha until recently has been known only from the type collection. Ducke in the original description recognized the affinity of this taxon to *H. intermedia* but distinguished *H. adenotricha* by (1) thickly coriaceous leaflets with the base broadly expanded on one side, (2) leaflets and petioles pubescent beneath, and (3) ovary with short, stiff hairs on one side. After examination of the type specimen and evaluating all characters, *H. adenotricha* is thought to be merely a western Amazonian variant of *H. intermedia,* occupying a geographical range distinct from that of the typical form (Lee and Langenheim, 1974a).

Ducke in the original description noted that the young fruit of this variety resemble those of *Peltogyne* and suggested a close relationship of *Hymenaea* and *Peltogyne.* This similarity, however, is only superficial and the relationship of the two genera is not, in fact, that close (Lee and Langenheim, 1974a).

11. *Hymenaea courbaril* L. Sp. Pl. 1192. 1753 (Figures 25, 26, 27)

Small to large tree, the bark smooth, silvery grey (rarely rough with deep fissure and grooves); the branchlets usually glabrous, rarely lightly villose. Leaves on slender or stout petioles, the petioles *ca.* 12-30 mm., glabrous or villose, the leaflets with twisted petiolule, 2-8 mm. long, the petiolule glabrous or lightly pubescent, the lamina falcate to obovate-oblong, 3-12 x 1.5-7 cm., glabrous and shining on both surfaces or lightly villose beneath, long- to short-acuminate, or obtuse, the base oblique, with the inner side acute and the outer side slightly to moderately rounded, the midrib and secondary veins usually prominent and elevated beneath, glabrous or densely villose. Inflorescence a short panicle, corymbose during anthesis, its branches glabrous to lightly ochraceous-puberulent, the pedicels *ca.* 3-10 mm. long. Flowers medium-sized (smaller in var. *altissima*); hypanthium campanulate, with a stalk-like base 4-8 mm. long, disc-bearing, usually with mass of additional nectariferous tissue immediate to the disc; calyx lobes obovate-oblong, obtuse, 12-22 x 7-18 mm., ochraceous puberulent to dark brown velutinous outside, yellowish to golden yellow sericeous inside; petals creamy white, obovate or rarely slightly spatulate, subsessile to short-clawed; stamens *ca.* 25-40 mm. long, the anthers *ca.* 3-8 mm. long; ovary subsessile to long-stipitate, obliquely oblong, glabrous throughout or rarely with some pilose hairs at one side at base, the style *ca.* 24-35 mm. long, the stigma capitate, prominent, the ovules 6-18(+). Fruit oblong to cylindrical, subterete to strongly compressed, brown to dark brown, shining, dotted with numerous resinous pockets and minute lenticels. Seeds 1-8(+) obovoid to ellipsoid, the testa light to dark brown, its intrusion into the cotyledons producing 2 shallow, horizontal furrows.

Key to Varieties of *Hymenaea courbaril*

1. Leaflets glabrous.
    2. Leaflets small (30-50 x 12-22 mm. in floral branches), distinctly falcate; flowers small (older buds *ca.* 15 x 8 mm.) . . . . B. var. *altissima*
    2. Leaflets medium-sized (*ca.* 50-120 x 30-70 mm.); broadly falcate to obovate-oblong; flowers medium-sized (older buds *ca.* 20-30 x 9-12 mm.).
        3. Calyx lobes ochraceous (usually pinkish before anthesis) tomentose on both sides.
            4. Leaflets obovate to oblong, rather large and broad (*ca.* 120 x 70 mm.) obtuse; petals obovate to slightly spatulate, sessile to subunguiculate (claws *ca.* 2 mm. long) . . . . C. var. *longifolia*
            4. Leaflets narrowly to broadly falcate, medium-sized (*ca.* 70-90 x 35-45 mm.), short- to long-acuminate, occasionally obtuse.
                5. Ovary with stipe 4-6(+) mm. long; leaflets broadly falcate, short-acuminate to obtuse, the base nearly straight to rounded on outer side; fruit large and oblong . . . . A. var. *courbaril*
                5. Ovary subsessile or with stipe *ca.* 2 mm. long; leaflets narrowly falcate, short- to long-acuminate, the base straight on outer side; fruit smaller, cylindrical . . . . E. var. *subsessilis*
        3. Calyx lobes golden brown to rusty velutinous outside, golden yellow inside . . . . D. var. *stilbocarpa*
1. Leaflets lightly to densely villous beneath . . . . F. var. *villosa*

Hymenaea courbaril occurs over a vast geographical area of tropical America and the Antilles. Only recently it is considered to be a polymorphic species consisting of six varieties (Lee and Langenheim, 1974a), one of them (var. *courbaril*) very widely distributed, the others more restricted in range. Further details are given under the respective varieties.

Hymenaea courbaril may be distinguished from other species of its section by its (1) medium-sized leaflets and flowers, (2) usually falcate, short-acuminate, glabrous leaflets, (3) hypanthium with a stalklike base *ca.* 6-8 mm. long, and (4) ovate, obovate, or rarely slightly spatulate petals.

11A. *Hymenaea courbaril* var. *courbaril* (Figures 25, 26)

Hymenaea courbaril var. *obtusifolia* Ducke, Arch. Jard. Bot. Rio de Janeiro, 4: 47. 1925. Based upon a cultivated tree in Museu Goeldi, Belém, Pará, Brazil (from Ilha do Marajó?), *Ducke 16906* (MG, holotype!; A!, P!, US!, isotypes).

Hymenaea multiflora Kleinhoonte, Recueil Trav. Bot. Neerl. 22: 405. 1925. Based upon *Herb. Boschwezen 4943*, Reserve Kaboeri, Surinam (U, holotype!).

Hymenaea resinifera Salisb. Prodr. 327. 1796.

Hymenaea animifera Stokes, Bot. Mat. Med. 2: 449. 1812, fide Index Kew.

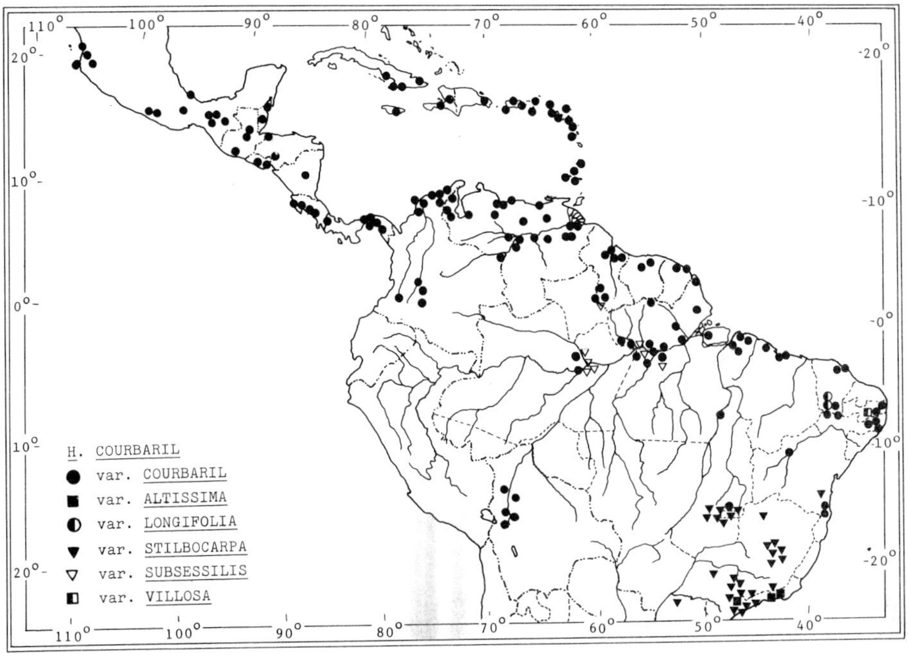

25. Distribution of *H. courbaril.*

Hymenaea candolleana Kunth, Nov. Gen. & Sp.6: 323, pl 556. 1824. Based upon *M. A. Bonpland 3875,* Acapulco, Mexico (P, holotype!; F!, isotype).
Hymenaea retusa Willd. ex. Hayne, Darst. Beschreib. Arzneigew. 11: pl. 12. 1856.
**Inga megacarpa* M. E. Jones, Contr. West Bot. 15: 140. 1929.

A large tree to 40 m. tall in forests, or more often much smaller in open or disturbed areas, the branchlets glabrous. Leaves short-petiolate, the petioles *ca.* 10-20 mm., glabrous, the leaflets with twisted petiolule 2-4 mm. long, the lamina falcate to broadly falcate, rarely oblong or obovate, 4-10 x 2-5 cm., coriaceous and shining on the two surfaces but duller beneath, short-acuminate to obtuse, rarely retuse, the base oblique, with the inner side acute and the outer side slightly to moderately rounded. The branches of inflorescence lightly ochraceous puberulent to glabrate. Pedicels *ca.* 6 mm. long. Hypanthium with stalklike base *ca.* 8 mm. long; calyx lobes ovate to elliptical, *ca.* 18 x 8 mm., ochraceous pubescent outside, yellowing sericeous inside; petals ovate to obovate, subsessile, *ca.* 20 x 9 mm.; stamens *ca.* 35 mm. long, the anthers *ca.* 6-8 mm. long; ovary short- to long-stipitate, glabrous throughout, *ca.* 4 x 2 mm., the stipe *ca.* 4-8 mm. long; the ovules 12-15(+). Fruit oblong to cylindrical *ca.* 8-14 x 4-6 x 1.5 cm. (occasionally reaching 20 x 8 x 3 cm.), light to dark brown, rounded at both ends, slightly to strongly compressed. Seeds 6-12 (+), obovoid or ellipsoid. Chromosome number 2N = 24.

Type: Description and illustration in L. Plukenet's Almagestum Botanicum, 2: 96, fig. 2. This is the type species of the genus.

In establishing the species *H. courbaril,* Linnaeus listed two references: the description and illustration by Plukenet (1696), and Bauhin's (1623) description of the species. Plukenet's description and illustration are chosen here as the lectotype of the species.

Habitat and distribution: Hymenaea courbaril var. *courbaril* has a range almost coincident with

---

*In "North American Flora" (Britton & Rose, 1930) this is listed as a synonym of *Hymenaea courbaril* L. According to Woodson & Schery (1951), Jones' description would preclude this possibility, but if Britton and Rose had reference to the actual type perhaps the description was incorrectly drawn by Jones. Description of the fruit matches *H. courbaril* but that of the leaves does not.

American taxa of the genus. It occupies all major tropical ecosystems ranging from rain forest to thorn forest (Table 2, Fig. 9e, f). It is, however, more frequent in the drier ecosystems both north and south of the Amazon Basin (Fig. 24) than in the *hylaea.*

Selected specimens: NORTH AMERICA. MEXICO. *J. A. Gaman 2* (US), *16* (US), VI-1961, *Langenheim 4707, 4788* (UC), X-1963, *4856, 4955* (UC), XI-1963, *4968, 4969* (UC), 27-XI-1898, *E. Langlasse 550* (A, F, K, P, US), 22-VI-1967, *Littman & Pennington 9037* (A, NY), 21-II-1937, *L. H. MacDaniels 152* (F), 16-V-1937, *E. Matuda 1426* (A, NY), 24-XI-1926, *Y. Mexia 1114* (A, F, NY, UC), 15-IV-1933, *Morton & Makrinius 2536* (K, US), 1895, *E. Palmer 330* (A, F, K, NY, U, UC, US), 19-IX-1967, *Pennington & Sarukhan 9217* (K), 1-III-1968, *9774* (K, NY), 10-IV-1910, *J. N. Rose et al. 14299* (NY, US), 26-VI-1897, *J. N. Rose 1489* (US), IV-1937, *L. Williams 9376* (F, K, NY, U, UC), VI-1937, *9567* (F).

CENTRAL AMERICA. BRITISH HONDURAS. 16-III-1931, *W. A. Schipp 741* (A, F, K, NY, UC), 14-VI-1933, *C. L. Lundell 4395* (UC, US), 10-11-1942, *P. H. Gentle 3920* (NY).

COSTA RICA. IV-1900, *A. Conduz 13967* (A, CR, F, K, P, US), 1-3-IV-1924 *Standley 40005* (US), 1931-1932, *W. Kupper 1502* (M), 1-III-1940, *A. C. Smith P2477* (A, UC), 6-VII-1949, *Holm & Iltis 302* (A), 15-IX-1963, *D. M. Porter 1137* (UC), 17-IV-1968, *J. J. Cordeba 909* (CR), 13-IV-1970, *R. Daubenmire 692* (F).

EL SALVADOR. 12-21-II-1922, *Standley 20659* (NY, US), 2-11-III-1922, *21695* (A, US), 7-14-II-1947, *Standley & Padilla 3705* (F), III-1923, *S. Calderon 1527* (NY, US).

GUATEMALA. 26-V-1914, *Cook & Doyle 91* (US), 2-VI-1966, *Jones & Facey 3525* (UC, US), 1-III-1907, *W. A. Kellerman 6571* (F), 27-IV-1933, *Lundell 3044* (F), 29-V-1933, *3459* (F), 16-IV-1966, *McClay & Clara 33* (A, F), IV-1893, *J. D. Smith 4469* (NY, US), 8-XII-1936, *Standley 59175* (F), 25-XI-1940, *78330* (F), 18-XII-1940, *80425* (F), 20-II-1941, *87841* (A), 20-II-1941, *87965* (F), III-IV-1941, *92677* (A, F), VI-1901, *H. von Tuerkeheim 7842* (A, F, M, NY, US), 19-V-1919, *Whitford & Stadtmiller 69* (US), *Leyde & Lee 4469* (M), *J. A. Steyermark 49190* (F, US, VEN).

HONDURAS. 26-V-1956, *A. Molina 7304* (F), 11-I-1948, *741* (F), 17-VI-1947, *323* (A, F, US), 24-IV-1949, *2378* (A, F), 23-VII-1947, *H. E. Agricola 378* (F), 5-16-III-1949, *Standley 17642* (F), 6-III-1931, *Schipp 741* (UC), 12-23-III-1947, *Standley & Chacon 5366* (F), 19-V-1919, *Whitford & Stadtmiller 57* (A, US).

NICARAGUA. 28-II-1903, *C. F. Baker 83* (NY, UC, US), *A. Garnier 3068* (F), 4-22-III-1928. *Smith 3109* (MG), V-VI-1947, *Standley 8802* (F), IV-V-1949 *20083* (F).

PANAMA. 8-II-1937, *P. H. Allen* (US), 14-V-1939, *1766* (US), 3-IV-1925, *O. F. Cook* (IPEAN), 16-V-1970, *T. B. Croat 10209* (NY, SU), 22-III-1971, *L. R. Holdridge 6434* (UP), 21-22-III-1911, *H. Pittier 3351* (A, NY, US), 4-XII-1923, *Standley 25866* (US), 12-I-1924, *30716* (US), 16-27-XI-1925, *40775* (US), 27-VII-1959, *Stern et al. 1229* (A, US), 31-V-1969, *Tyson & Lazor 5334* (SU), 31-V-1969, *5335* (SU, UC), 31-V-1969, *5336* (SU), 21-VI-1969, *Tyson 5559* (SU).

WEST INDIES. ANTIGUA. 15-VII-1938, *H. E. Box 1506* (A, US), 4-16-II-1913, *J. N. Rose et al. 3491* (A, NY, US), *C. Gallo 583* (NY).

CUBA. XII-1928, *Clement 2300* (A, NY, US).

DOMINICA. 7-VII-1931, *E. J. Valeur 685* (A), 8-20-II-1923, *W. L. Abott 2609* (US), 8-20-II-1923, *2619* (US), 20-VII-1964, *B. Augusto 1670* (NY).

GRENADA. 2-VI-1909, *W. E. Broadway* (A, NY), 7-25-III-1950, *R. A. Howard 10865* (A).

GUADELOUPE. 1880, *P. Duss 3044* (US), 12-VII-1951, *C. L. Gallo* (NY).

HAITI REPUBLIC. 30-V-1926, *E. L. Ekman 6197* (US).

JAMAICA. 29-VIII-1907, *N. L. Britton 949* (NY), 22-VIII-1962, *Proctor 22627* (U), 7-XI-1959, *19811* (A), 1844, *Hooker 49* (NY), 6-VII-1896, *E. Campbell 6414* (F, NY).

MARGARITA. *J. R. Johnston 290* (A), *A. Plée* (P).

MARTINIQUE. 13-I-1900, *O. Penzig* (K), 1874, *M. Hahn 906* (A, NY, US), 3-VII-1939, *F. E. Egler* (NY), 2-VII-1901, *Underwood & Griggs 573* (US).

MONTSERRAT. 30-I-1959, *Proctor 18935* (A, US), 6-II-1907, *Shafer 392* (F, NY, US), *J. S. Beard 433* (K).

NETHERLANDS ANTILLES. 1945, *M. Arnoldo 1950* (U), 30-VII-1906, *I. Boldingh 2332* (U), 18-VIII-1906, *2592* (U), 28-VIII-1906, *2881* (U).

PUERTO RICO. 8-IX-1967, *T. G. Hartley 13382* (U), 6-VI-1968, *R. J. Wagner 1565* (A, U), 29-IV-1968, *1485* (A, U), 13-VI-1885, *I. Urban 1521* (K, US), 7-VII-1931, *E. J. Valeur 685* (A, K, US), 8-VII-1950, *E. L. Little 13280* (NY, D), 27-IV-1887, *P. Sintensis 6759* (A, K).

SANTO DOMINGO. 13-VII-1930, *E. L. Ekman 15303* (A, F, K, NY, US), I-III-1871, *C. Wright et al.* (A, US), 13-VI-1930, *H. Harms 1935* (K).

ST. CROIX. 31-VIII-1896, *A. E. Ricksecker 340* (A, F, NY, UC, US), 26-II-1897, *430* (F, US), 1870, *L. Hahn 222* (B), 1906, *C. Raunkiaer 2530* (P, US).

ST. EUSTATIUS. 17-VII-1906, *Boldingh 1764 (889)* (U), 17-VIII-1953, *A. L. Stoffers 4735* (U), *J. V. Groll-Meyer 216* (U), 22-I-1932, *D. Fairchild 2731* (A, US).

ST. LUCIA. IV-VI-1958, *Proctor 17817* (A, US), IV-V-1950, *Howard 11461* (US).

ST. THOMAS. I-1919, *H. M. Curran 806* (A), 30-III-1896, *Ricksecker 340* (A, UC), 5-7-II-1913, *Britton & Shafer 325* (NY, UC), 7-X-1959 *Proctor 19811* (A, NY).

ST. VINCENT. III-1889, *H. H. & G. W. Smith 195* (A, B, NY), 4-IV-1947, *C. V. Morton 4727* (A, US), 16-IV-1947, *5127* (A, US), *E. H. L. Krause 11816* (B).

TOBAGO. 18-IV-1910, *Broadway 3534* (F, NY).

TORTOLA. 13-17-II-1913, *Britton & Shafer 933* (NY, US).

TRINIDAD. 20-IV-1928, *Broadway 6894* (F, K), *E. Eppes US 429597* (US).

SOUTH AMERICA. BOLIVIA. 8-I-1902, *R. S. Williams 595* (NY), 1901-1902, *586* (NY), 7-IX-1922, *G. E. White 1211* (F, NY), 26-IX-1965, *H. Garcia-Barriga 18549* (NY), 1-22-VII-1939, *Krukoff 10255* (A, F, U).

COLOMBIA. XI-1898/1901, *H. H. Smith 2027* (A, F, NY, P, UC, US, UT), 11-I-1941, *Dugand & Jaramillo 2760* (COL, US), I-1948, *R. Romero-Castaneda 673* (COL), 14-I-1952, *Inst. de Parcelaciones 60* (COL), 21-IV-1960, *L. U. Uribe 3455* (COL), I-1937, *A. G. Barriga 5155* (COL), I-1856, *J. Triana 6785* (COL), 1931, *Espina 19* (F), VII-1920, *M. T. Dave 966* (K), 30-III-1934, *A. Dugand 555* (F), 28-II-1903, *C. F. Baker 83* (SP).

FRENCH GUIANA. 14-VII-1921, *Broadway 824* (A, NY, US), XI-XII-1956, *afog 1219* (U), *M. Melinin 18* (K, R), V-1856, *Sagot 894* (K, NY, P), IV-1856, *Sagot* (K, NY).

GUYANA. 3-12-IX-1922, *J. S. de la Cruz 1931* (A, F), 28-VII-1920, *L. S. Hohenkerk 845* (K), 6-VIII-1886, *Jenman 2242* (K), 8-X-1958, *R. Persand 1615* (K, NY), 9-15-XI-1937, *A. C. Smith 2485* (A, F, NY, P, U, US), 4-22-III-1938, *3109* (A, F, IPEAN, NY, U, US), XII-1948, *G. Wilson-Browne 65* (K, NY).

SURINAM. 18-IX-1920, *Boschwezen 4943* (U), 5-V-1922, *5819* (U), 31-V-1917, *2899* (U), 26-VII-1918, *3925* (U), 7-VII-1922, *5917* (IPEAN, U, UB, US), 5-V-1922, *HERB Suriname 5819* (IPEAN, NY), 18-IX-1963, *H. S. Irwin et al. 55843* (A, COL, NY, U, UC, US), 28-V-1924, *J. C. Huyzen 6417* (U, US), 12-VIII-1950, *Bosbeheer 1098* (U), IX-1945, *G. Stahel 367* (A, K, NY).

VENEZUELA. 28-V-1914, *Steyermark 56768* (F), 18-V-1967, *R. A. Pursell et al.* (NY), 1939, *P. C. Vogel 1341* (M), 29-IV-1955, *Wurdack & Wright 408* (F, NY, US, VEN), 17-IV-1940, *L. Williams 12860* (F, US, VEN), 25-IV-1939, *11899* (UC, US, VEN), 20-IV-1955, *Tamayo 4024* (US, VEN), 5-V-1950, *F. D. Smith 168* (US), 8-IV-1950, *131* (US), IV-1941, *J. Saer 800* (US), V-1896, *Rusby & Squires 205* (A, F, M, NY), V-1896, *204* (A, F, M, NY), 25-IV-1937, *H. Pittier 13998* (F, NY, US, VEN), 18-IX-1969, *Langenheim 5592* (UC), 15-IX-1969, *5591* (UC), 2-VI-1939, *E. Delgado 252* (US, VEN), 13-V-1940, *Chardon 58* (US, VEN), 17-II-1946, *A. Burkart 16052* (VEN), IV-1967, *J. De Bruijn 1725* (VEN), 19-V-1964, *L. M. Berti 169* (F, NY, US, VEN), 9-IX-1966, *C. Blanco 606* (NY, US, VEN).

BRAZIL. Pará. Belém and its vicinity. IX-1936, *Ducke 277* (A, F, K, R, US); Museu Goeldi, 20-VIII-1910, *J. Huber 10860* (MG), VIII-1922 *Ducke 16906* (K, MG, US), IX-1936, *278* (A, F, NY, US), 11-VIII-1942, *W. A. Archer 7559* (F, NY, US), 13-II-1946, *J. J. Kubling* (RB), 8-VIII-1957, *Cavalcante 286* (MG), 28-IX-1969, *Langenheim 5594* (UC); Highway Belém-Brasilia, km. 94, 29-IX-1959, *Kuhlmann & Jimbo 314* (G, INPA, M, SP, UB), km. 162-170, 10-V-1960, *Oliveira 729* (IPEAN); Rio Jaracu, 9-X-1955, *R. L. Fróes 32170* (IPEAN); Rio Piracicaba, 28-VIII-1957, *Fróes 33381* (IPEAN, MG); Mosqueiro, 22-XI-1971, *Lee 75, 76* (UC); Faro, 26-I-1911, *Ducke 10548, 10549* (MG), 1-II-1910, *10626* (MG); Obidos, 26-VII-1912, *Ducke 12051* (MG), 10-III-1915, *15714* (MG), Santarém, Palhão Reserve, 8-X-1969, *Langenheim 5621* (INPA, UC); Curuá-Una, *Langenheim 5624* (INPA, UC); 24-XII-1963, *Rodrigues & Freitas INPA 14273* (INPA); Bragança-Vizeu, X-1966, *J. T. Santos 7792* (MG); Trombetas, 18-I-1968, *Ducke* (MG); Monte Alegre, 12-II-1933, *Fróes 29530* (IPEAN), 7-V-1953, *Andrade-Lima 53-1418* (IPA, IPEAN); Rio Pará do Oeste, 12-II-1970, *Cavalcante 2359* (MG, NY); Rio Jarí, 30-X-1968, *Silva 1339* (IPEAN), 27-IV-1968, *Oliveira 4341* (MG); Taruma-Miry, 22-XII-1912, *Ducke 12406* (MG); Araguaia, 27-VI-1953, *Fróes 22926* (MG); Marajó, 5-X-1899, *Guedes 1938* (MG).

Maranhão. Boa Esperanca, Maracassumé River, 18-VIII-1932, *Fróes 1803* (A, F, LP, NY, P): São Luiz, II-III-1939, *Fróes 11527* (LP), *11528* (F, K, NY, U, UC, US).

Amazonas. Manaus and its vicinity. 30-III-1959, *J. Chagas 7223* (US), 24-VIII-1955, *Rodrigues INPA 1703* (INPA), VI-1966 *INPA 17266* (INPA), 23-VII-1962, *Rodrigues & Chagas 4552* (INPA), Rio Purus, X-1917, *Ducke 16824* (MG); Beruri, 7-IV-1967, *Silva 818* (MG, NY).

Bahia. Barra and its vicinity. 10-X-1912, *Zehntner 330* (RB); Santo Antonio, 22-X-1912, *Zehntner 439* (R); Rio Grongogy Basin, X-XI-1915, *H. M. Curran 73* (A); Banuro, 26-X-1912, *Zehntner 6301* (R); 88 km. from Vitoria de Conquista, 25-X-1969, *Langenheim 5640* (UC).

Paraíba. Near Perari, 29-XI-1971, *Lee 84, 85, 86, 87, 88, 89* (UC); near Monte Horebe, 30-XI-1971, *91, 92.*

Ceará. Quixadá, 14-IV-1909, *Ducke 2165* (MG); Fortaleza, 12-X-1937, *T. N. Guedes 586* (IPEAN, (MG); Serra da Baturité, 5-XII-1937, *B. I. Azevedo RB44693* (RB); Alcantara, 15-IV-1954. *Fróes 30789* (IPEAN, R); Messejana-Eusebio, 10-XI-1956, *Ducke 2575* (IPEAN, MG); 11 km. S. of Barro, 1-XII-1971, *Lee 93, 94* (UC); 8 km. W. of Crato, 2-XII-1971, *Lee 96* (UC); Chapada do Araripe, 2-XII-1971, *Lee 97, 100, 101* (UC); Caririzinho, 3-XII-1971, *Lee 107* (UC); Mun. de Vicosa, Assemi, 22-VI-1972, *D. Sucre 9250* (RB, UC).

Piauí. Serra da Capivara, 6-XII-1971, *Lee 123* (UC).

Mato Grosso. Urupas, I-1919, *Kuhlmann 2011* (R, RB, SP); Serra do Roncado, Xavantina, 31-V-1966, *H. S. Irwin 16308* (NY).

Goías. Goiania, XII-1935, *Brade 15424* (RB); Rio Piranha Araguatins, 28-IV-1961, *Oliveira* (IPEAN). Brasilia, D. F., 4-XI-1969, *Langenheim 5655* (UC).

Terr. Amapá. Serra do Navio, 18-XI-1971, *Lee 71* (UC).

Vernacular names:* "Courbaril, locust" (South America, West Indies), "algarobo" (Spanish America), "guapinol" (Central America), "Coapinol, cuapinole, pak, nere, quauhpinolli" (Mexico), "simiri, locust" (Guyana), "semirie, locus" (Surinam), "corobore, algarobo" (Venezuela), "algarobo, nazareno (Colombia), "algarobo, jatahy, jutahy, jatai, jatai assu, jatai grande, jataí, jatobá, jataíba, trapuca, yatayba, yutahy" (Brazil), "abati, timbary, avati" (Paraguay), "asucar-huain, pampa estoraque" (Peru).

11B. *Hymenaea courbaril* var. *altissima* (Ducke) Lee & Lang. J. Arnold Arbor. 55(3): 448. 1974 (Figures 25, 26)

Hymenaea altrissima Ducke, Ann. Acad. Brasil Sci. 7: 207. 1935.

Large emergent trees to 40 m. tall in forests, or usually much smaller in highly disturbed areas. Leaves on slender petioles, the petioles 12-15 mm. long, glabrous, the leaflets with twisted petiolule 3-5 mm. long, the lamina of the leaflets on floral branches falcate, 3-5 x 1.2-2.2 cm., much larger on vegetative branches, short- to long-acuminate, the base oblique, with the inner side acute and the outer side slightly rounded, the midrib and secondary veins prominent on both surfaces, the reticulate pattern of veinlets evident above. Inflorescence branches rather short, densely subgolden brown puberulent, the pedicels *ca.* 3 mm. long, bracts and bracteoles *ca.* 7 x 5 mm.. caducous, before anthesis the developing buds golden brown velutinous. Flowers rarely more than 15 mm. long, hypanthium with stalk-like base *ca.* 4 mm. long; calyx lobes 12-15 x 7-10 mm., brownish rust to golden yellow sericeous, more densely inside than outside; petals obovate to oblanceolate, sessile, 16-18 x 6-8 mm.; stamens *ca.* 26 mm. long, the anthers 3-4 mm. long; ovary moderately long-stipitate, glabrous throughout, the stipe *ca.* 2-4 mm. long, the style *ca.* 24 mm. long. Fruit variable in size, the 1-seeded *ca.* 4.0-5.0 x 2.5-3.0 cm., those 2- to 6-seeded up to 12 x 50 x 20 mm. Seeds obovoid to ellipsoid. Chromosome number 2N = 24.

Type: Brazil. State of Rio de Janeiro, Avellar, Faz. Posse. 1930. *A. Ducke RB 30306* (RB, lectotype!; MG!, R!, isotypes). Since the original author merely listed six collections but did not indicate a type, the first of the six was chosen as the type by Egler (1963).

Habitat and distribution: This subtropical variety has a narrow distribution along the Atlantic Coast. It has been collected in humid forest in the city of Rio de Janeiro, Guanabara, in Serra do Mar of the State Rio de Janeiro, and the city of São Paulo. Trees bud in November, flower in December, and have mature fruit in the following May.

Specimens examined: BRAZIL. Guanabara. Rio de Janeiro and vicinity: XI-1879, *M. Glaziou*

---

*For less commonly used names refer to Record & Mell (1924) and Record and Hess (1954).

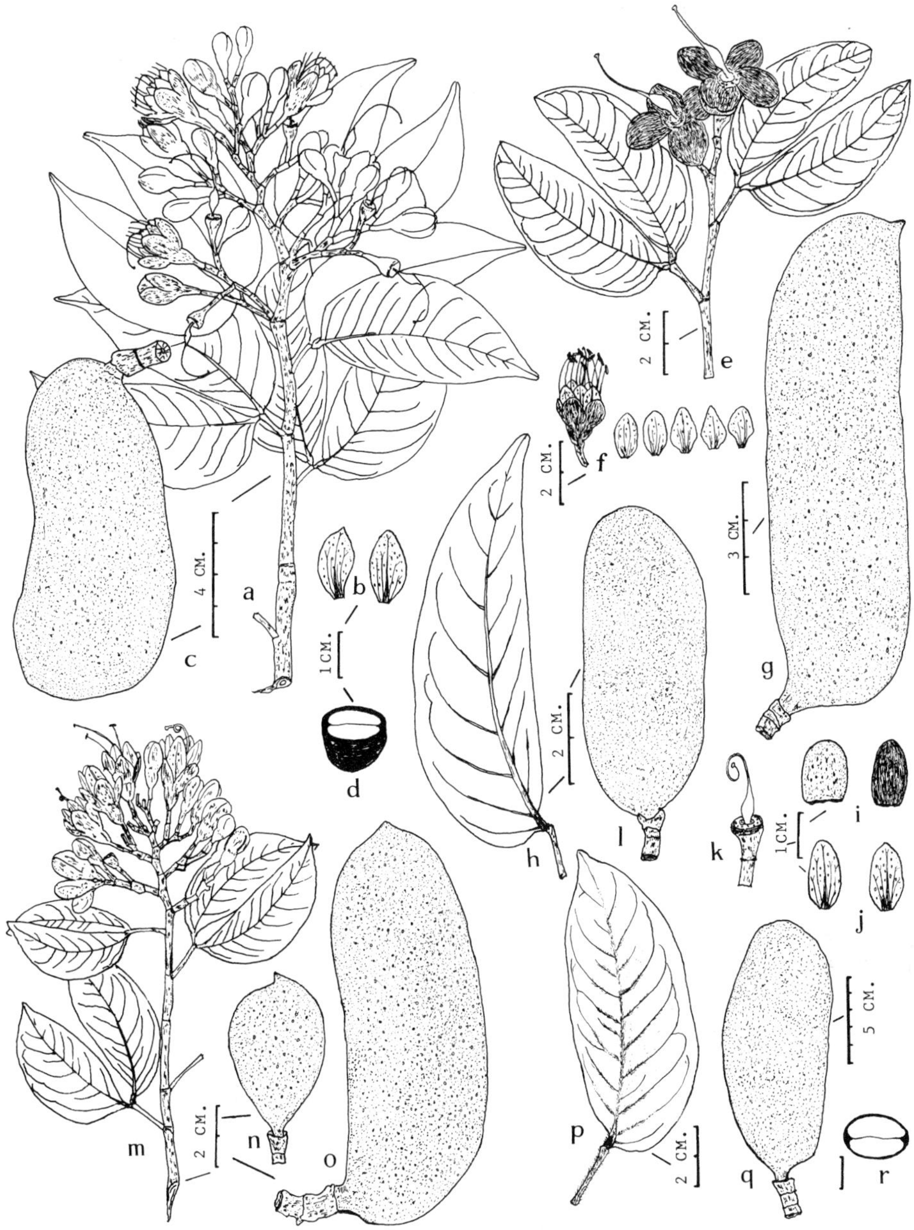

26. *Hymenaea courbaril* (excluding var. *latifolia*). a-d, var. *courbaril:* a. flowering branch. b. petals. c. fruit. d. seed (cross-section). e-g, var. *stilbocarpa:* e. flowering branch. f. flower and petals. g. fruit. h-1, var. *subsessilis:* h. leaflet (adaxial side). i. sepals. j. petals k. pistil, hypanthium, and pedicel. l. fruit. m-o, var. *altissima:* m. flowering branch. n. one-seeded fruit. o. many-seeded fruit. p-r, var. *villosa:* p. leaflet (abaxial side). q. fruit. r. seed (cross-section): a-d, g, after Hayne (1830); e and f, after Mattos (1968); m-o, after Ducke (1935).

10697 (K), 26-XI-1927, *RB 112639* (RB), 6-XII-1927 *Kuhlmann 21486* (P, RB) 2-XII-1926, *21453* (P), 30-X-1952, *J. de Saldanha 37* (R).

Rio de Janeiro. XI-1925, Avellar, *G. M. Nunes* (RB), 1935, *Servico de Reflorestamento* (R).

São Paulo. São Paulo city and vicinity: 4-XII-1931, *F. C. Hoehne 28566* (F, P, RB, SP), 24-28-XI-69, *Langenheim 5667, 5673* (UC), 2-I-1972, *Lee 130, 133, 134, 135* (UC).

Goías. Near Niquelândia. 23-I-1972, *Lee 159* (UC).

Vernacular name: "Jatahy," which is commonly used for the entire genus in southern Brazil.

Hymenaea altissima is characterized by its (1) small, short- or long-acuminate, and distinctly falcate leaflets on floral branches, (2) highly reticulate venation conspicuous on both surfaces, (3) relatively small flowers, and (4) one- or many-seeded fruit.

The relatively small and falcate leaflets on the floral branches and relative size of flowers do not appear sufficiently significant to warrant species status for this taxon. Of the two types of fruits described by Ducke, the one-seeded fruits are ovoid to obovoid and subcompressed, whereas the many-seeded fruits are oblong and strongly compressed, essentially similar to those of *H. courbaril* var. *courbaril.* Thus we concluded that the relationship of *H. altissima* is best expressed by varietal status under *H. courbaril* (Lee and Langenheim, 1974a); the occurrence of two different types of fruits may be the result of factors controlling the availability of pollen during anthesis.

A tree growing along a river near Niquelândia, Goías *(Lee 159)* has leaflets very similar to those of var. *altissima* but without floral material, identification is impossible.

11C. *Hymenaea courbaril* var. *longifolia* (Benth.) Lee & Andrade-Lima, J. Arnold. Arb. 55(3): 448. 1974 (Figures 25, 27)

Hymenaea splendida var. *longifolia* Benth. in Mart. Fl. Bras. 15(2): 237. 1870.

Small to medium-sized trees 6-12 m. tall, the bark smooth, silvery-grey, or rough with deep fissures and grooves at base of trunk, the branchlets glabrous, rusty brown. Leaves long-petiolate, the petioles *ca.* 15-30 mm. long, glabrous; the leaflets with twisted petiolule *ca.* 8 mm. long, the lamina obovate-oblong, 8-12 x 4-7 cm., glabrous and shining, obtuse to short-acuminate, the base oblique, with the inner side acute and the outer side moderately rounded, the midrib and secondary veins prominent on both surfaces, elevated beneath, the veinlets obscure. Inflorescence a short panicle, loosely corymbose, composed of a few racemes to 10 cm. long, its branches glabrate, light brown, longer than in other varieties, the pedicels 8-10 mm. long. Flower medium-sized (older buds *ca.* 30 x 12 mm.); hypanthium campanulate, with a stalklike base *ca.* 8 mm. long, disc prominent, with additional mass of nectariferous tissue below; calyx lobes obovate-oblong, obtuse, *ca.* 22 x 18 mm., ochraceous puberulent outside, yellowish tomentose inside; petals subequal, obovate to slightly spatulate, subsessile to unguiculate (claws to 3 mm. long); stamens *ca.* 40 mm. long, the anthers *ca.* 8 mm. long; ovary long-stipitate, *ca.* 12 x 4 mm., broader at apex, glabrate or with some pilosity on one side at base, the stipe *ca.* 8-10 mm. long, the style *ca.* 35 mm. long, the ovules 12-18(+). Fruit light brown, dotted with numerous, small, resinous pockets, 15 cm. long, 5.5 cm wide (broader toward apex, narrower at base), 4.2 cm. thick, oblong to rectangular, subcompressed, pointed at distal end. Seeds not seen.

Type: Bahia. Villa de Barra, *Blanchet 3135* (R, holotype!, P!, isotype).

Habitat and distribution: Other than the type collection this variety has been collected in Chapada do Araripe in the border area of the states of Ceará, Piauí, and Pernambuco, surrounded by *caatinga* characteristic of northeastern Brazil. The vegetation of the *chapada* or highlands is relatively dense, with scattered medium-sized trees, and a dense ground cover (*cerrado* or *cerradão*). *Hymenaea courbaril* var. *longifolia* flowers in November and December. Fruit falls about September.

Specimens examined: BRAZIL. Ceará, Chapado do Araripe, 3-XII-1971, *Lee & Andrade-Lima, 95, 110, 111, 112* (UC).

This taxon was originally described by Bentham as a variety of *H. splendida* and has been

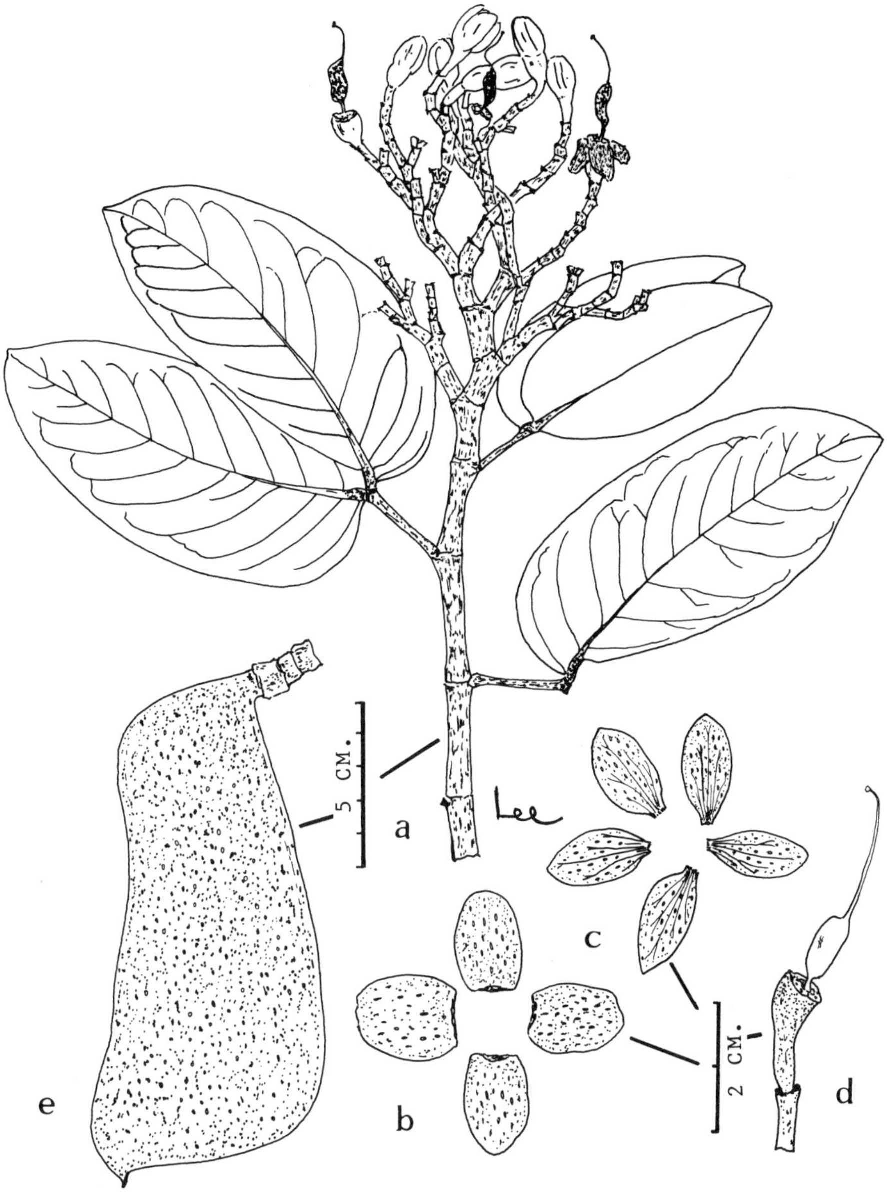

27. *Hymenaea courbaril* var. *longifolia*. a. flowering branch. b. sepals. c. petals. d. pistil, hypanthium, and pedicel. e. mature fruit.

known only from the type collection. Additional specimens were collected during a trip to northeastern Brazil, which subsequently enabled us to determine its present taxonomic status (Lee and Langenheim, 1974a). *Hymenaea courbaril* var. *longifolia* is characterized by (1) large, rather long, and usually obtuse leaflets, (2) medium-sized flowers, (3) occasionally clawed petals, (4) externally ochraceous puberulent, and internally yellow tomentose calyx, and (5) unusual shape of the fruit, which is broadest at its distal end.

This variety may be separated from *H. stigonocarpa* var. *stigonocarpa,* which is characterized by glabrate leaflets and is commonly found in the *cerrado* of central Brazil, by its smaller flowers, long-stalked hypanthium (the stalklike base is 6-7 mm. long), and the shape of the fruit.

11D. *Hymenaea courbaril* var. *stilbocarpa* (Hayne) Lee & Lang. J. Arnold Arbor. 55(3): 449. 1974 (Figures 25, 26)

Hymenaea stilbocarpa Hayne, Arzneik. gebräuchl. Gewächses 11: pl. 11. 1830.
Hymenaea confertifolia Hayne, Arzneik. gebräuchl. Gewächses 11: pl. 9. 1830. Based upon *Sellow* (B?, holotype).
Hymenaea splendida Vogel, Linnaea 11: 409. 1837. Based upon *Sellow 1025* (B?, holotype, photographs, A!, F!, NY!, US!).

Trees 12-20 m. tall, the trunk to 65 cm. in diameter, the branchlets glabrous. Leaves short-petiolate, the petioles 12-20 mm. long, the leaflets with twisted petiolule 4-6 mm. long, the lamina falcate, 5-7.5 x 2.5-3.5 cm., coriaceous, glabrous and shining, acute to short-acuminate, the base oblique, with the inner side acute and the outer side rounded, the midrib and secondary veins barely evident below, veinlet reticulate above. Inflorescence composed of few racemes, each few-flowered, its branches and pedicels rusty sericeous-puberulent, the pedicels 6-10 mm. long. Flowers medium-sized, the buds blackish to dark brown, velutinous; hypanthium with a stalklike base *ca.* 4-9 mm. long; calyx lobes *ca.* 12-18 x 8-12 mm., orbiculate to broadly ovate, densely dark brown to rusty sericeous or occasionally golden brown velutinous outside and golden brown to golden yellow sericeous inside; petals ovate to obovate, *ca.* 18 x 8 mm., sessile; stamens *ca.* 32-35 mm. long, the anthers *ca.* 6-7 mm. long; ovary short-stipitate, glabrous throughout, the stipe *ca.* 3-4 mm. long, the ovules 10-12(+). Fruit very oblong to cylindrical, obtuse at both ends, subterete to slightly compressed, brown to dark brown, occasionally black and shining. Seeds 1-6(+), ellipsoid, 20-25 x 15-18 x 10-12 mm., the testa dark brown. Chromosome number 2N = 24.

"Representative": Brazil. "Ad Caldas in Minas Gerais," IX-1867, *Regnell 11-91* (K!). Hayne (1827, 1830) cited no specimens in his original description of this taxon. One of the four collections listed by Bentham (1870), the first set of specimens ever cited under this taxon, has been chosen as the representative specimen (Lee and Langenheim, 1974a).

Habitat and distribution: This variety is native along rivers in relatively dry forests of the plateau of the state of São Paulo and adjacent parts of Rio de Janeiro and Minas Gerais. It has also been collected in southern Bahia, in the Federal District, and southern Goías. It is frequently cultivated in Belo Horizonte, Minas Gerais. It flowers from November to January; the fruit is usually mature in September.

Specimens examined: BRAZIL. F. D. Parque Nacional de Brasilia, 4-XI-1969, *Langenheim 5656* (UC), 17-VI-1970, *S. G. Fonseca 1614* (RB).

Minas Gerais. Município Santa Luzia. 2-X-1942, *J. Nestor* (IPEAN); Lagoa Santa, 16-I-1951, *Andrade-Lima 51-794;* Paraopeba, 20-X-1959, *SP 79892* (SP), 25-VIII-1957, *Fróes 33245* (IPEAN), 20-X-1971, *RB 1053583* (RB); Municipio de Belo Horizonte, 11-X-1941, *Nestor* (IPEAN); 9-I-1935, *M. Barreto 173* (K, P, RB, U, US), 2-X-1941, *11221* (RB), 11-I-1935, *174* (F, R), *RB 5174* (RB), Ramal de Montes Claros, IX-1929, *RB 25433* (K, RB, US), *Schwacke 13396* (RB), *Claussen 794* (F, P).

Goiás. Rio Corumbá, 30 km. from Luziânia, 3-XI-1969, *Langenheim 5652* (UC); Niquelândia, I-1972, *Lee 156, 157* (UC).

São Paulo. Pindorama. 10-IX-1937, *O. T. Mendes 2137* (RB); Moji-Mirim, 18-XI-1936, *SP 36863* (SP), 22-I-1972, *Lee 136, 137, 138, 139* (UC); Moji-Gauçu, *Mattos 10565* (SP); 26-XI-1969, *Langenheim 5669, 5670* (UC); São Paulo City and vicinity, 27-XI-1957, *RB 102643* (RB), 25-28-XI-1969, *Langenheim 5668, 5672* (UC), 12-IV-1931, *Hoehne 28566* (A); Limeira, XI-1931, *J. G. Kuhlmann 20529* (K, U, US); Piracicaba, 20-X-1956, *Herringer 5394* (UB), 20-X-1959, *Herringer 9420* (NY, UB).

Bahia. Rio Grande, *Zehntner 1924* (RB); Vitoria, 25-X-1969, *Langenheim 5641, 5642* (UC).

Paraná. Rio Paranapanema, Municipio S. Antonio do Cairra, 22-VI-1966, *G. Hatschbach 14450* (NY, SI, US).

Vernacular names: "Jatobá, jatobá amarelo" (Minas Gerais), "jatobá mirim" (São Paulo), "jatobá miudo" (Ceará), "jatobá de caatinga" (Bahia).

In describing this taxon Hayne (1830) emphasized its dark brown sausagelike fruit. Although it has been confused with var. *altissima,* it is usually more like var. *courbaril,* and Bentham stated (1870) that fruit of the two could scarcely be distinguished. Ducke (1935) predicted that *H. stilbocarpa* would eventually be considered a subtropical variety of *H. courbaril.* The leaflets of var. *stilbocarpa* resemble those of certain populations in Mexico and the West Indies in shape, texture, and shiny surface, a similarity possibly explained by progressive radiation into adjacent subtropical regions. After evaluating all the characters, we agree with Ducke's view that the taxon is best considered as a variety of *H. courbaril* (Lee and Langenheim, 1974a).

Hymenaea courbaril var. *stilbocarpa* is characterized by (1) calyx lobes being rusty or dark brown sericeous-velutinous externally, golden sericeous internally, (2) fruit often dark brown and occasionally subterete, (3) flowers smaller than in var. *courbaril* and *subsessilis* but much larger than in var. *altissima,* and (4) medium-sized, subfalcate leaflets.

11E. *Hymenaea courbaril* var. *subsessilis* Ducke, Arch. Jard. Bot. Rio de Janeiro, 4: 263. 1925 (Figures 25, 26)

Medium-sized to large trees, occasionally much smaller along sandy river beaches. Leaves long-petiolate, the petioles 16-22 mm. long, the leaflets with twisted petiolule 2-4 mm. long, the lamina falcate, 8.5-10 x 2.5-3.5 cm. glabrous and shining, coriaceous, short- to long-acuminate, the base oblique, with the inner side acute and the outer side slightly rounded, the midrib prominent below, the secondary vein and veinlets obsolete on both surfaces. Inflorescence a short panicle, densely corymbose during anthesis, many flowered, the pediçels *ca.* 6 mm. long. Flowers medium-sized, the developing buds ochraceous-green, usually pinkish; hypanthium campanulate, subsessile; calyx lobes 14-18 x 10-14 mm., subequal, lightly and subequally ochraceous tomentose on both sides; petals subequal, ovate to obovate, *ca.* 12-15 x 7-9 mm., sessile; stamens *ca.* 30 mm. long, the anthers *ca.* 6-8 mm. long; ovary short-stipitate or more often subsessile, the style *ca.* 20 mm. long, the ovules 8-12 or more. Fruit cylindrical, very short-stalked, subcompressed, the suture subevident. Seeds 4-6(+), ovoid to ellipsoid, the testa dark to light brown. Chromosome number 2N = 24.

"Representative": Brazil. Amazonas. Manaus, Flores, 18-XI-1910, *A. Ducke MG 11167* (MG!). The selection of a representative specimen is necessary because no type specimen was designated by the original author. This particular specimen was chosen because (1) it was collected in the general area believed to be the type locality, and (2) it is complete, having foliar and floral material as characterized for this taxon.

Habitat and distribution: The variety *subsessilis* is more frequent in both the eastern and central Amazon Basin than var. *courbaril* and occurs in a somewhat wetter habitat, such as sandy river beaches. It is an emergent tree when found on *terra firme* but appears in much reduced size elsewhere. The flowering season is from October to November. Fruit falls about August.

Specimens examined: BRAZIL. Pará: Rio Tapajós, I-1918, *Ducke 17017* (R); near Periquito, *11099* (RB); Santarém, 6-X-1969, *Langenheim & Rodrigues 5629, 5631* (INPA, UC).

Amazonas. Rio Jamundá, Santa Rosa, 16-XI-1950. *Black & Ledoux 50-801* (COL, INPA, IPEAN, MG, R); Aminaru-Assu, 18-XI-1950, *50-10867* (IPEAN); Rio Negro. Ilha do Camalião, 1-X-1969, *Langenheim & Rodrigues 5611* (INPA, UC); Manaus and its vicinity, 21-IV-1882, *Schwacke 457* (R), 2-X-1969, *Langenheim & Rodrigues 5612, 5613* (INPA, UC), 24-VIII-1955, *INPA 1703* (INPA, UB), 20-II-1958, *INPA 6075, 6076* (INPA), 23-VII-1962, *Rodrigues & Chagas 4552A* (INPA), 18-VIII-1970, *Rodrigues 8905* (INPA), 17-IV-1966, *Rodrigues, Langenheim, & Andrades-Lima 7906* (INPA), VIII-IX, 1936, *Krukoff 7952* (A, F, K, P, U, US), 30-XI-1931, *Ducke 24282* (K, P, RB, U, US), 1-I-1941, *Ducke 667* (F, IPEAN, MG, R, UC, US), 20-26-X-1971, *Lee 49* (UC), 1-XI-1971, *62* (UC).

Hymenaea courbaril var. *subsessilis* is characterized by (1) usually falcate and acuminate leaflets, (2) nearly sessile ovary, (3) subcylindrical fruit, and (4) the ochraceous green tomentum which occurs about equally on both sides of the calyx. The specimens collected from *campos* near Almeirim, Pará, immediately upstream from the estuary, according to Ducke (1935) appear morphologically intermediate between vars. *subsessilis* and *courbaril.*

11F. *Hymenaea courbaril* var. *villosa* Lee & Andrade-Lima, J. Arnold Arbor. 55(3): 446. 1974 (Figures 25, 26)

Medium-sized tree 8-10 m. tall, the trunk *ca.* 35 cm. in diameter, the branchlets lightly puberulent to to glabrate. Leaves short-petiolate, the petioles slender, 12-18 mm. long, lightly to densely pubescent, the leaflets with twisted petiolule *ca.* 5 mm. long, the petiolule pubescent, the lamina broadly to narrowly falcate, 6-9 x 3.5 cm., coriaceous, shining and glabrate above, lightly to densely tomentose-villose beneath, occasionally conspicuously short-hairy along margins, acute to short-acuminate, the base oblique, with the inner side acute and the outer side slightly rounded, the midrib and secondary veins elevated beneath and lightly to densely villose, the veinlets finely reticulate above. Inflorescence branches, pedicels, and flower buds lightly ochraceous to yellowish brown puberulent, the pedicels *ca.* 6 mm. long. Flowers relatively small; hypanthium with a stalklike base *ca.* 6-7 mm. long; calyx lobes obovate, *ca.* 14 x 7 mm., ochraceous to subgolden brown pubescent, densely golden brown tomentose inside; petals obovate, obtuse, *ca.* 12 x 7 mm., subsessile; stamens *ca.* 22-25 mm. long, the anthers *ca.* 5-6 mm. long; ovary short-stipitate, oblique-oblong, relatively small, *ca.* 3 x 2 mm., slightly pilose at one side of base, less often glabrous, the stipe *ca.* 3 mm. long, the ovules 8-12(+). Fruit oblong to cylindrical, *ca.* 9-12 x 4.5-5.5 x 3.5-4.0 cm., dark brown, strongly compressed, the suture subevident. Seeds 4-8(+), obovoid to ellipsoid, *ca.* 2.5 x 2.0 x 1.2 cm. Chromosome number 2N = 24.

Type: Brazil. Paraíba: Along roadside near Paquevira de Natuba, 27-X-1971, *Lee & Andrade-Lima 83* (UC, holotype!).

Habitat and distribution: The type specimens were collected from a moist forest in hills (elevation *ca.* 500 m.) in inland Paraíba about 70 km. west of the Atlantic coast. Other collections were made in Paraíba.

Specimens examined: BRAZIL. Paraíba. In the vicinity of Natuba. 11-27-X-1971, *Lee & Andrade-Lima 81, 82* (UC); Alagoinha, 4-I-1943, *L. Xavier 1183* (Herbarium, Unversidade Federal da Paraíba).

Pernambuco. Tapera, 6-XII-1931, *D. B. Pickel 2853* (A, IPA).

Vernacular name: "Jatobá."

This taxon is restricted to hills in eastern Paraíba. Morphologically it is very close to var. *subsessilis* but may be distinguished from it by (1) the light to densely villose lower surface of the leaflets, particularly the midrib, (2) the smaller flowers, and (3) the occasional pilosity at the inner side of the ovary base.

The occasionally very densely hairy leaflets have caused this variety to be confused with *H. martiana,* which is generally found in the *caatinga* or interior areas of Brazil. Variety *villosa* may be distinguished from *H. martiana* by its broadly to narrowly falcate leaflets, smaller flowers, and vesture of calyx.

12. *Hymenaea martiana* Hayne, Arzneik. gebräuchl. Gewächses 11: pl. 15, 1830 (Figures 8g, 28)

Hymenaea sellowiana Hayne, Arzneik. gebräuchl. Gewächses 11: pl. 16, 1830. Based on *Sellow,* "Brazilia" (K, holotype!).

Medium-sized or large trees to 15 m. tall, the young branchlets pubescent. Leaves short-petiolate, the petioles 12-18 mm. long, densely pubescent, the leaflets with twisted petiolule 4-6 mm. long, the

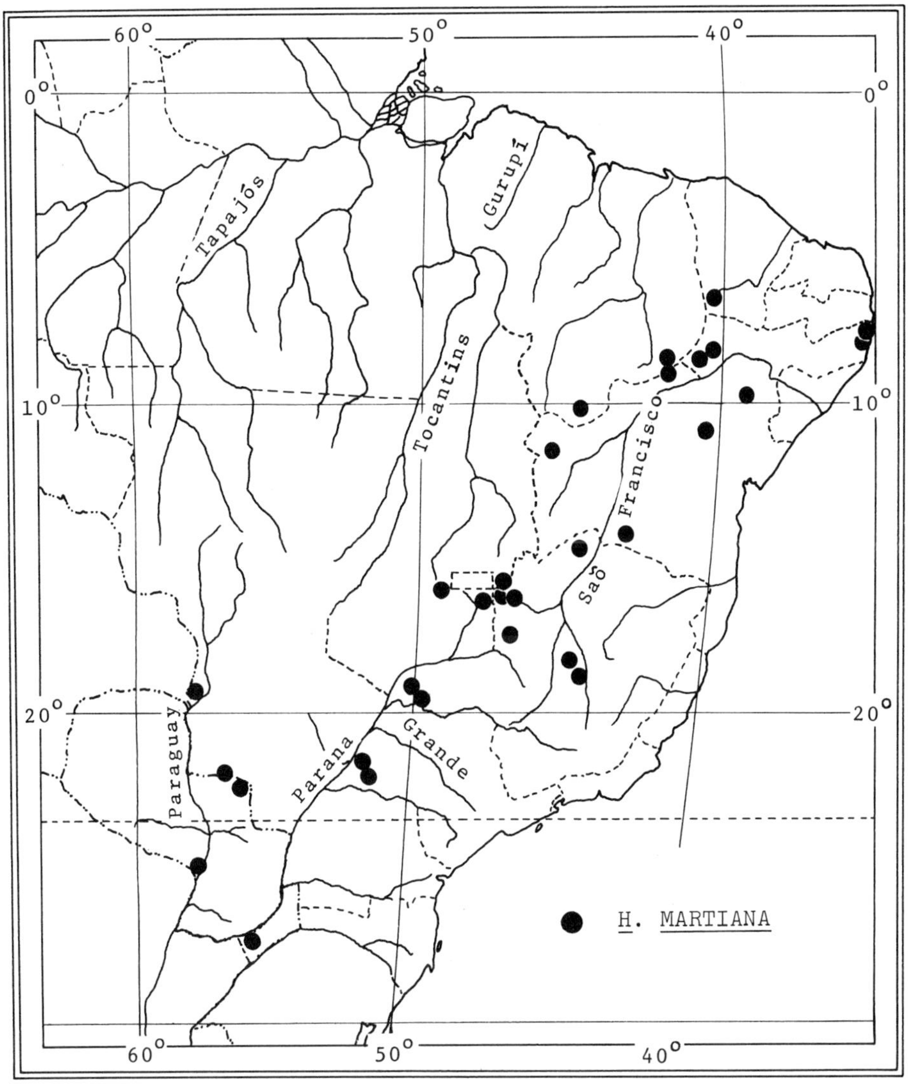

28. Distribution of *H. martiana.*

petiolule pubescent, the lamina ovate to obovate-oblong, 5.5-7.5 x 2.5-4.5 cm., the adaxial surface of adult leaflets glabrate and shining, sparsely subappressed pilose, the abaxial surface densely greyish tomentose, coriaceous, obtuse, the base oblique, with the inner side acute and the outer side moderately rounded, the midrib and secondary veins obscure above, densely tomentose and prominent beneath. Inflorescence a short panicle, densely corymbose during anthesis, many-flowered, its branches thickly greyish tomentose; the pedicels 4-7 mm. long, bracts and bracteoles broadly ovate, caducous. Flower medium-sized, hypanthium campanulate, with a stalklike base 4-6 mm. long, disc prominent, with additional mass of nectariferous tissue below; calyx lobes ovate to obovate, *ca.* 12-14 x 8-10 mm., ochraceous tomentose outside, whitish sericeous inside; petals ovate-obovate, sessile, subequal, *ca.* 15 x 6-8 mm., creamy white; stamens *ca.* 25 mm. long, the anthers *ca.* 6 mm. long; ovary long-stipitate, usually with a tuft of hairs at base, less commonly glabrous, the stipe *ca.* 6 mm. long, the style *ca.* 18 mm. long, the ovules 8-12. Fruit oblong, compressed-cylindrical, very smooth, sometimes with markedly raised suture, shining, blackish brown to brown, *ca.* 9-12 x 3-4.5 x 2.5-3.5 cm. Seeds 4-6(+), obovoid to ellipsoid, the testa dark brown, its intrusion into cotyledons producing 2 shallow, horizontal furrows. Chromosome number 2N = 24.

"Representative": Brazil. "ad Paranagoa, Prov. Piauhy." VIII-1839, *Gardner 2533* (K!). Since Hayne did not cite any specimen in the original description, one of the five collections listed by Bentham (1870) is chosen here as the representative specimen.

Habitat and distribution: Hymenaea martiana occupies drier habitats than does *H. courbaril,* and it is commonly found as a medium-sized tree in savannah *(cerrado)* in Goiás and northern Minas Gerais. However, it extends north into the thorn forests *(caatinga)* of the four northeastern states, Piauí, Ceará, Pernambuco, and Bahia, and southward to Paraguay and Argentina. Although it grows in relatively dry ecosystems, it is most frequent along river banks. The flowering season is from October to January. Fruit falls about August.

Specimens examined: BRAZIL. Alagoas. União dos Palmares, 26-XI-1966, *Andrade-Lima 66-4744* (IPA).

Bahia. Villa Nova (NE Bahia), XI-1913, *Luetzelburg 70* (M), 725 (NY); Barra, 10-X-1912, *Zehntner 331* (RB); Caetetê, 22-XI-1912, *608* (M, RB); Barreiro, Rio Corrente, 9-XI-1912, *504* (RB); on road to São Raimundo Nonato, *ca.* 5 km. S of the border of Piauí, 6-XII-1971, *Lee 124* (UC); Jacobina, 1839, *J. Blanchet 2648* (K).

Ceará. Juazeiro, 24-I-1911, *Loefgren 829* (R).

Goiás. 1896, *A. Glaziou* (K), *Burchell* (K); near Luziânia, Rio Corumbá, 5-XI-1969, *Langenheim, 5658, 5659, 5660, 5661* (UC); Rodovia Brasilia-Pires do rio margem do Rio Corumbá, 25-XI-1966, *Herringer 11224* (NY, UB); Serra Dourada, 30 km. SE Goiás Velho, 22-I-1966, *Irwin, et al.* (UB); Rod. Luziânia a Vianopolis, 15-X-1968, *Herringer 11854* (NY, UB); Ilha do Catalão, D. F., 1960, *R. Delforge* (NY).

Mato Grosso. Saloba, 30-X-1952, *A. Santos* (R); 25-X-1912, *Zehntner 460* (R); Corumbá, 15-XII-1950, *Barton* (RB).

Minas Gerais. Paraopeba, 31-XI-1960, *Herringer 7779* (UB); Rod. Brasilia, Rio Abaté, 31-XI-1960 *779* (NY); Famosa 10-XI-1951, *M. Magalhães 4581* (IPEAN); Ituitutaba, 14-X-1944, *S. R. Morais 553,* 14-X-1946, *A. Macedo 840* (SI), 27-X-1948, *1300* (SI); Rio das Velhas, 29-VII-1929, *G. M. Nunes 50;* Paracatu, 20-VI-1964, *Pires 58052* (NY, UB).

Pernambuco. Paulista, 29-VII-1950, *Andrade-Lima 50-760* (IPA, IPEAN); Buique, 10-III-1955, *55-2022* (IPA); Cabrobó, 13-XI-1958, *58-3298* (IPA); Olinda, 25-I-1924, *852* (IPA); Parnamirim, 6-XII-1971, *Lee 125* (UC).

PARAGUAY. X-1930, *P. Jorgensen 3763* (A, F, K, NY, SI, US), X-1907, *Hassler 9717* (A), 1904, *7792* (A, F, K, NY, P, UC).

ARGENTINA. Candelaria, X-1945, *J. Barriero* (SI).

Vernacular name: "Jatobá," which is also commonly used for other *Hymenaea* species in southern Brazil.

This species exhibits characters similar to those of *H. courbaril,* particularly var. *villosa.* It is distinguishable by (1) ovate to obovate leaflets, which are obtuse and densely greyish-white tomentose beneath, (2) the ovary usually with a tuft of hair at base, (3) the calyx densely greyish white tomentose externally, and (4) a stalklike hypanthium base *ca.* 6 mm. long.

It shows considerable similarity to *H. velutina* and *H. stigonocarpa,* but can be separated from the former by its ovate to obovate, densely tomentose leaflets and the larger and strongly compressed fruits, and from the latter by its smaller flowers, the long-stalked hypanthium, and the size, form, and pubescence of leaflets.

These four named species form a coherent group within sect. *Hymenaea* characterized by (1) the well-developed disc with additional nectariferous tissue within the hypanthium, (2) ovate to obovate petals, (3) large flowers, and (4) oblong or cylindrical fruits.

Hymenaea martiana can be separated from *H. reticulata* and *H. rubriflora* by its small, ovate to obovate leaflets and ovate to obovate, subsessile petals; and from *H. eriogyne, H. aurea,* and *H. maranhensis* by its glabrate ovary and rather small, ovate to obovate leaflets.

13. *Hymenaea velutina* Ducke, Arch. Jard. Bot. Rio de Janeiro 4: 48. 1925 (Figures 8f, 29, 30)

Small tree, 4-6 m. tall, the crown broadly conical, the bark smooth, silvery grey. Leaves glabrous throughout, the petiole slender, 1.5-2.5 cm. long, the leaflets with twisted petiolule *ca.* 3 mm. long, the lamina obovate-oblong, 4-8.5 x 2-4.5, glabrous, coriaceous, obtuse to retuse, the base oblique, with the inner side acute to slightly rounded and the outer side broadly expanded and strongly rounded, the midrib and secondary veins prominent on both surfaces, the veinlets finely reticulate, more prominent above. Inflorescence a short panicle, densely corymbose during anthesis, its branches densely reddish to dark brown puberulent to velutinous, the pedicels 6-8 mm. long. Flowers medium-sized, older buds oblong, 3.0-3.5 x 1.0-1.2 cm., hypanthium campanulate, with a stalklike base *ca.* 8 mm. long; calyx lobes oblong-obovate, obtuse, 18-22 x 9-13 mm., dark brown to blackish golden brown velutinous outside, golden yellow tomentose inside; petals creamy white, obovate, 20-25 x 13-18 mm., sessile, subequal; stamens *ca.* 25 mm. long, the anthers 6 mm. long; ovary glabrous throughout, long-stipitate, the stipe *ca.* 6 mm. long, the style *ca.* 18 mm long, the ovules 11-15(+). Fruit usually light brown, oblong to cylindrical, slightly compressed, 8-11 x 4-5 x 3-4 cm., with the broad faces marked by conspicuous subdiagonal veins, the suture not prominent. Seeds 3-6(+), ellipsoid to ovoid, 18-22 x 16-20 x 8-10 mm., the testa light brown, its intrusion into the cotyledons producing 2 shallow, longitudinal furrows. Chromosome number 2N = 24.

Type: Brazil. Piauí: "Habitat in civitatis Piauhy regionis Piracuruca campis Taboleiro Grande," 7-X-1909 (RB, lectotype!; MG!, US!, isotypes). Ducke listed three collections without designating a type. A lectotype is chosen here from these three specimens.

Habitat and distribution: Hymenaea velutina is restricted to northeastern Brazil. It is commonly found in small tree thickets *(cerradão)* and at the margins of outcrops of rocks in thorn-forest area in southern Piauí, Ceará, Maranhão, and northern Bahia. The flowering season is from November to December. Fruit falls about October.

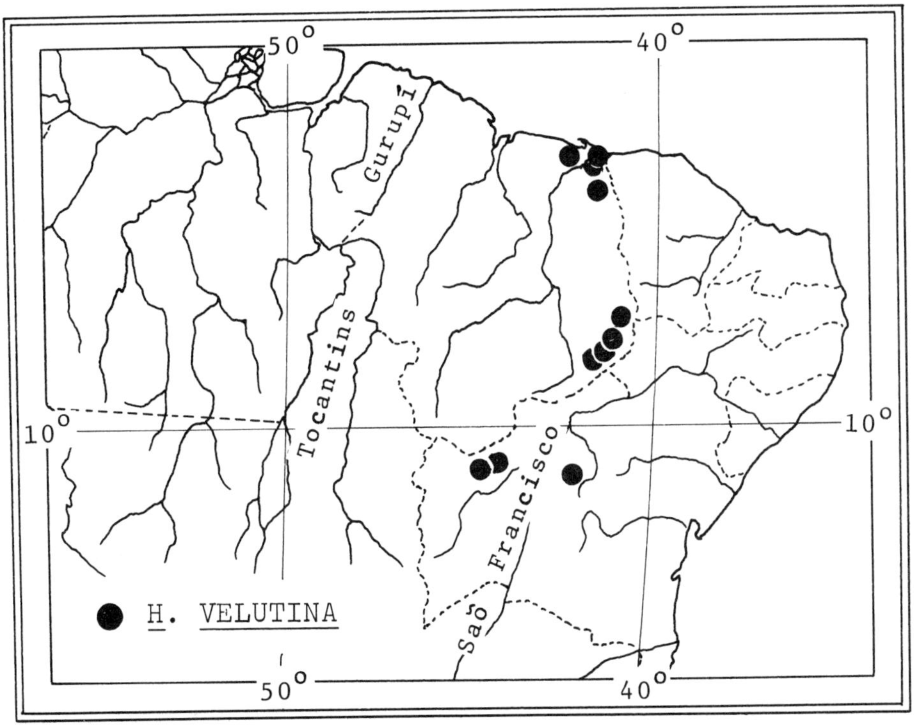

29. Distribution of *H. velutina.*

Specimens examined: BRAZIL. Piauí: Parnaíba, 13-VII-1954, *Andrade-Lima 54-2091* (IPA); Canta Galo, 22-VI-1972, *Sucre 9276* (RB, UC); Highway BR-230, S E of Picos, 5-XII-1971, *Lee & Andrade-Lima 114, 115* (UC); 14 km. SW of Itainopolia, *118* (UC); 16 km. S of Simplicio Mendes, 6-XII-1971, *120* (UC).

Bahia. Logo abaixo de Santa Rita. IX-1912, *Zehntner 375, 377* (R).

Maranhão. Salinas Tutoia, 17-VII-1937, *Ducke 16910* (RB).

Vernacular names: "Jatobá de casca fina" (southern Piauí); "jatobazinho" and "jatobá da caatinga" (northern Bahia).

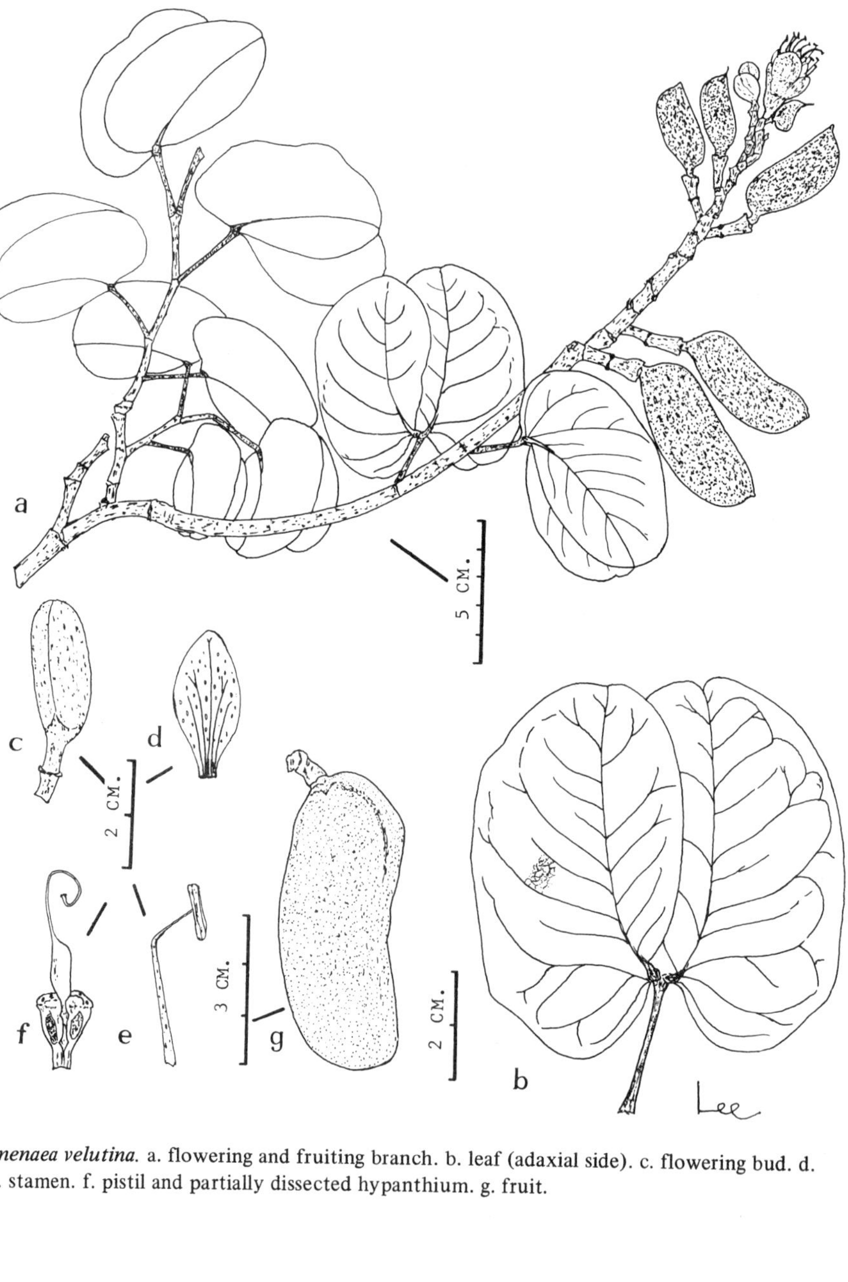

30. *Hymenaea velutina.* a. flowering and fruiting branch. b. leaf (adaxial side). c. flowering bud. d. petal. e. stamen. f. pistil and partially dissected hypanthium. g. fruit.

Hymenaea velutina has its greatest affinity with *H. courbaril,* particularly var. *longifolia.* It can be separated from *H. courbaril* by its (1) leaf base broadly prolonged on one side, (2) obovate-oblong leaflets, and (3) rather small, narrow, and slightly compressed fruits. A close relationship with *H. martiana* and *H. stigonocarpa* can also be demonstrated by (1) presence of a well-developed hypanthial nectary, (2) ovate or obovate, sessile petals, (3) a ± glabrous ovary, and (4) long epicotyls in the seedling. It can be separated from *H. martiana* by its glabrous leaflets and slightly compressed fruits, and from *H. stigonocarpa* by its glabrous leaflets, smaller flowers, and slender-stalked hypanthium.

Although *H. velutina* has been observed growing sympatrically with *H. eriogyne* in northeastern Brazil (Fig. 8f), it can be separated from the latter by its (1) glabrous ovary, (2) unique leaf shape, and (3) seedlings with a rather long (2+ cm.) epicotyl.

14. *Hymenaea stigonocarpa* Mart. ex Hayne, Arzneik. gebräuchl. Gewächses 11: pl. 13. 1830 (Figures 31, 32)

Small to medium-sized evergreen tree 5-12 m. tall, the bark very rough, deeply fissured and grooved, or rarely smooth and silvery grey, the branchlets rusty, the epidermis usually peeling. Leaves short- to long-petiolate, the petioles stout to slender, 6-30(+) mm. long, the leaflets with twisted petiolule 3-8 mm. long. the petiolule glabrate to pubescent, the lamina generally very large (*ca.* 24 x 16 cm. or more), thickly coriaceous, glabrate to densely pubescent on both surfaces, obtuse or short-acuminate, the base oblique, with the inner side acute to rounded and the outer side moderately to strongly rounded, the midrib and secondary veins usually obscure above but prominent beneath, glabrate to densely hispid-hirsute. Inflorescence a short panicle, composed of few racemes, each relatively few-flowered, its branches glabrate to densely whitish grey puberulent, usually pinkish and covered with small pieces of peeling epidermis, the bracteoles relatively large, persistent almost to anthesis or early deciduous. Flowers largest for the genus (older buds up to 26 x 17 mm.); the hypanthium broadly obconical, subsessile, the disc very prominent, with additional mass of nectariferous tissue below; calyx lobes ovate to obovate, whitish grey or pale rusty tomentose outside, greyish tomentose-sericeous inside; petals ovate-obovate, glabrous, subsessile; stamens *ca.* 30-45 mm. long, the anthers 8-10 mm. long; ovary stipitate, glabrate, or very rarely with a few white marginal hairs, green when young, turning red upon aging, the ovules 10-18(+). Fruit oblong to cylindrical, rather narrow, slightly compressed, brown to dark brown, the sutures very prominent. Seeds 6-12(+), ellipsoid to obovoid, compressed or not, the testa light to dark brown, its intrusion into the cotyledons producing 2 longitudinal, shallow furrows.

Key to Varieties of *Hymenaea stigonocarpa*

1. Leaflets glabrate on both surfaces or lightly puberulent beneath when young; petioles 1.5-3.5(+) cm. long, glabrate . . . . A. var. *stigonocarpa*
1. Leaflets densely pubescent beneath.
    2. Petioles 1.5-4.5(+) cm. long, densely tomentose; leaflets 12-24 x 8-16 cm.; midrib and secondary veins densely tomentose beneath . . . . C. var. *pubescens*
    2. Petioles *ca.* 5-6 mm. long, densely woolly-tomentose; leaflets *ca.* 10 x 6 cm.; midrib and secondary veins densely hispid-hirsute beneath. . . . . . . B. var. *brevipetiolata*

Hymenaea stigonocarpa is notable among all species of the genus for its (1) large leaflets and flowers, (2) sessile or subsessile hypanthium, (3) large nectariferous disc, (4) densely greyish white woolly calyx, and (5) distinctive growth form which is shrubby or never more than a medium-sized tree, with the trunk often gnarled and twisted and the bark rough.

This species is closely related to *H. courbaril,* particularly var. *longifolia,* but the two are clearly separable by the characters listed above. A close relationship to *H. martiana* and *H. velutina* is indicated by the several characters they share. The similarity in certain

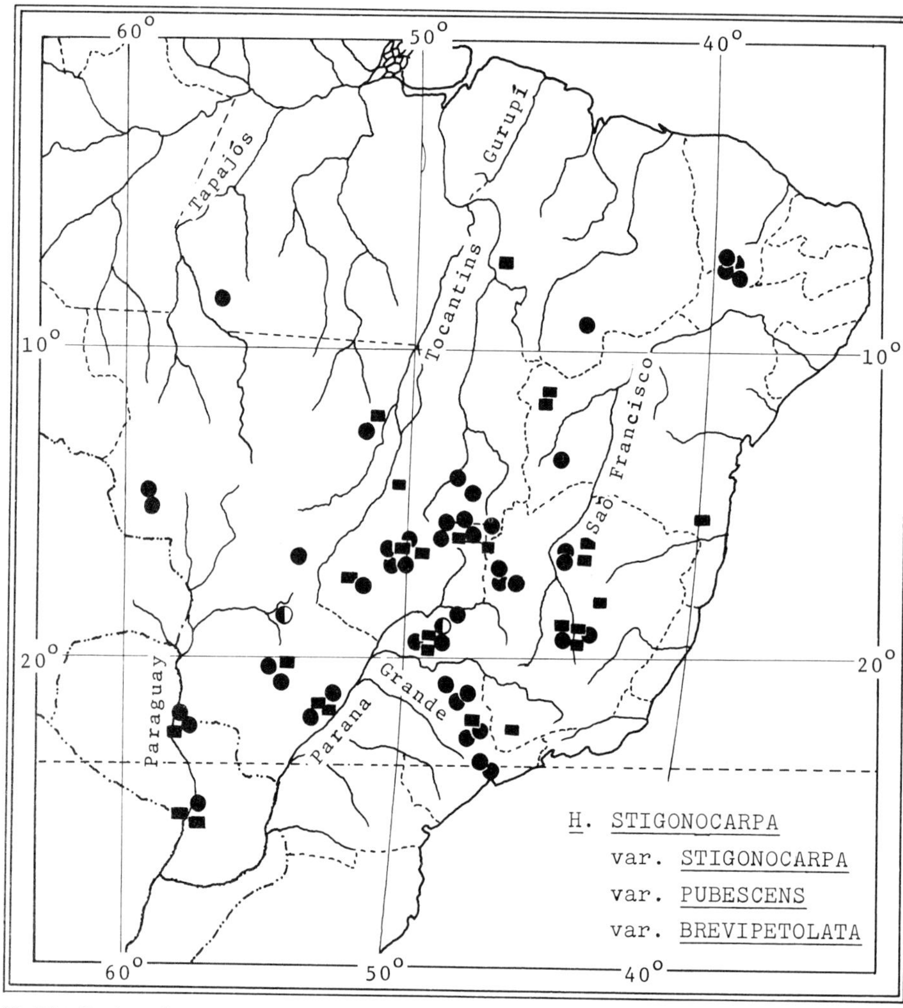

31. Distribution of *H. stigonocarpa.*

floral and foliar characters between *H. stigonocarpa* and *H. maranhensis* is also noteworthy, although they can be separated easily by pubescence of the ovary and shape of petals.

14A. *Hymenaea stigonocarpa* var. *stigonocarpa* (Figures 31, 32)

Small tree *ca.* 5 m. tall, the bark very rough, or occasionally smooth and silvery grey, the branchlets glabrate. Leaves long-petiolate, the petiole glabrous, 2-3 cm. long, the leaflets with twisted petiolule *ca.* 6 mm. long, the lamina obovate-oblong, occasionally slightly falcate, coriaceous, glabrate or lightly puberulent on both surfaces when young, obtuse, occasionally acute, the base oblique, with the inner side acute and the outer side moderately to strongly rounded, the midrib and secondary veins prominent, elevated beneath. Inflorescence branches short, lightly puberulent, pinkish, the pedicels *ca.* 6 mm. long. Flower large (older buds *ca.* 26 x 17 mm.), hypanthium broadly obconical, subsessile to sessile disc and additional nectariferous tissue prominent; calyx lobes whitish grey tomentose outside, greyish tomentose-sericeous inside; petals *ca.* 22 x 12 mm., sessile; stamens *ca.* 38 mm. long, the anthers *ca.* 9

mm. long; ovary stipitate, glabrous or occasionally with a few white marginal hairs, the stipe *ca.* 6-8 mm. long, the ovules 10-18(+). Fruit oblong to cylindrical *ca.* 12 x 4.5 x 3.5 cm., the suture prominent to subevident, usually dark brown, very smooth, with no apparent resinous pockets. Seeds 6-8(+), the testa light to dark brown, its intrusion into the cotyledons producing 2 shallow, horizontal furrows.

"Representative": Brazil. Minas Gerais. Cachoeira do Campos(?), 1839, *P. Claussen & B. Delerrert 300* (K!). Since Hayne (1830) cited no specimen in the original description, selection of a representative specimen is made here. That chosen here is from the five specimens listed by Bentham (1870).

Habitat and distribution: The typical variety shares its distribution pattern with var. *pubescens* in interior Brazil, extending north to Piauí and south to Paraguay. It is less common than var. *pubescens* although the two usually grow sympatrically in Goiás and Mato Grosso. This variety flowers from January to March. Fruit falls about December.

Specimens examined: BRAZIL. Bahia. Valley of the Rio das Ondas, 3-4-III-1971, *Irwin et al. 31369, 31478* (NY).

Goiás. 26 km. W of Rio Verde on road to Caiaponia, 3-II-1959, *Irwin 2560* (NY, R, UC); 78 km. E of Aragarcas on road to Piranhas, 21-VI-1966, *Irwin et al. 17510* (NY); Valley of Rio Maranhão, 20-I-1968, *18920;* 26 km. NE of Catalão, 23-I-1970, *25224* (NY); Dois Irmãos, 20-I-1972, *Irwin et al. 34607* (NY, UC); Brasilia, D. F., vicinity of Paranoá, 13-XII-1965, *11255* (NY), 1-I-1972, *Lee 150* (UC).

Maranhão. Porto de Carolina, 26-V-1950, *Pires & Black 2209* (IPEAN).

Minas Gerais. Contagem, Betim, II-1945, *L. O. Williams 5167* (A); Serra do Cipó. W of Morro do Pilar, 13-II-1959, *Irwin 2451* (NY, R, US); Uberaba, 1849, *A. F. Regnall III-489* (K, R, US); Corinto, 21-IV-1942, *M. Magalhães 1984* (IPEAN, US); Belo Horizonte, 6-III-1945, *Williams* (F, R, SP); Paraopeba, 24-VI-1966, *W. Nascimento* (SP, US); Campo Florido, 13-II-1964, *Labouriau & Valio 1162* (SP); Alfenas, 1-III-1969, *J. P. P. Carauta 820* (US).

Mato Grosso. 12-VI-1946, *H. Veloso 1341, 1347,* (RB); Xavantina and its vicinity, *Philcox & Fereira 3819* (NY), 1-VI-1966, *Hunt 5713* (NY), 7-VI-1966, *Hunt & Ramos* (K, NY, SP), 12-VI-1966, *Irwin et al. 17019* (NY); Campo Grande, 29-I-1969, *Granta 781* (US); Serra do Roncador, 94 km. N of Xavantina, 4-VI-1966, *Irwin et al. 16569* (NY).

São Paulo. Moji-Gauçu, 19-IX-1956, *O. Handro 626* (SP).

PARAGUAY. 24-VII-1875, *B. Balansa 1485* (K), *Hassler 1848* (K).

Vernacular names: "Jatobá" and "jatobá do cerrado" (Brazil).

This variety is distinguished from *H. courbaril* var. *longifolia* by its (1) few-flowered inflorescence which is composed of 4-8 short racemes, (2) the larger flowers, (3) the subsessile hypanthium, and (4) the growth form characteristic of the species.

14B. *Hymenaea stigonocarpa* var. *brevipetiolata* N. Mattos, Loefgrenia 16: 1-2. 1964 (Figures 31, 32)

Tall and multi-branched trees ("Abtore alta e frondosa"). Leaves short-petiolate, the petioles stout, *ca.* 6 mm. long, 3 mm. wide, densely whitish tomentose, the leaflets with twisted petiolule *ca.* 4 mm. long, the lamina obovate-oblong, relatively small, *ca.* 10 x 6 cm. (smaller in floral branches), coriaceous, thinly pubescent above and densely tomentose beneath, obtuse to retuse, the base oblique, with the inner side acute and the outer side moderately rounded, the midrib and secondary veins prominent, lightly tomentose above and hispid-hirsute beneath. Inflorescence branches whitish grey, densely woolly tomentose, the pedicels *ca.* 6 mm. long. Flowers smaller than in other varieties; hypanthium broadly obconical with a stalklike base *ca.* 2 mm. long; calyx lobes densely white woolly tomentose outside, petals not seen; stamens *ca.* 35 mm. long, the anthers *ca.* 6 mm. long; ovary long-stipitate, the stripe *ca.* 8 mm. long. Fruit and seeds not seen.

Type: Brazil. Minas Gerais: Ituiutaba, Dec. 14, 1946, *A. Macedo 842* (SP, holotype!; MG!, RB!, UC!, isotypes).

Habitat and distribution: Except for the type collection, this variety has been collected only once, in Mato Grosso (Serra da Chapada), but it may have a wider distribution in *cerrado* as do the other varieties. The flowering season is December.

Specimen examined: BRAZIL. Mato Grosso: Serra da Chapada, *Riedel* (A!).

This variety is distinguished by (1) short, thick petioles rarely exceeding 6 mm., (2) the dense whitish tomentose pubescence on petioles, rachides, and calyx, and (3) the smaller, obovate leaflets.

14C. *Hymenaea stigonocarpa* var. *pubescens* Benth. in Mart. Fl. Bras. 15(2): 236. 1870 (Figures 8h, 31, 32)

Hymenaea olfersiana Hayne, Arzneik. gebräuchl. Gewächses 11: pl. 14. 1830. *H. stigonocarpa* var. *olfersiana* O. Ktze. Rev. Gen. 3(2): 65. 1898.
Hymenaea rotundata Hayne, Arzneik. gebräuchl. Gewächses 11: pl. 13. 1830.
Hymenaea chapadensis B. Rodr. Pl. Mattogross. 23. – Bras., pl. 7, fig. A. 1898.
Hymenaea correnana B. Rodr. Pl. Mattogross. 21. – Bras., pl. 8. 1898.

Small to medium-sized trees 6-9 m. tall, the branchlets lightly rusty pubescent, the epidermis usually peeling. Leaves long-petiolate, the petiole *ca.* 3.5-4.5 cm. long, densely pubescent, the leaflets with twisted petiolule 5-8 mm. long, the lamina large and broad (the largest for the genus), obovate to oblong, occasionally rounded, 18-24 x 10-16 cm., occasionally even larger, thickly coriaceous, thinly pubescent above and densely pubescent beneath, obtuse, rounded or retuse, the base oblique with the inner side slightly to moderately rounded and the outer side moderately to strongly rounded, the midrib and secondary veins prominent, elevated and densely tomentose beneath, veinlets obscure. Inflorescence composed of a few short racemes, each few-flowered, the branches whitish grey tomentose, bracteoles relatively large, *ca.* 25 x 15 mm., persistent almost to anthesis. Flowers large (older buds *ca.* 35 x 18 mm.); hypanthium broadly conical, sessile to subsessile; stamens with filament *ca.* 35 mm. long, the anthers 8-9 mm. long; ovary long-stipitate, glabrous, or occasionally with a few marginal or basal hairs, the stipe *ca.* 8 mm. long, the ovules 10-18(+). Fruit oblong, rather narrow, *ca.* 12 x 3.5 x 3.0 cm., pointed at end, the suture elevated and prominent. Seeds 6-8(+), ellipsoid to obovoid, occasionally much compressed, the testa light brown to brown. Chromosome number 2N = 24.

Type: Brazil. Piauí, "A small tree common in chapadas near Oeiras," IV-1839, *Gardner 2158* (K, lectotype!; P!, isotype). Of several specimens cited in the original description, *Gardner 2158* is chosen here as the lectotype.

Habitat and distribution: This variety, more frequent than var. *stigonocarpa,* is abundant in *cerrado* and *cerradão* regions of northeastern and central Brazil. It flowers from December to March. Fruit falls about January.

Specimens examined: BRAZIL. Bahia: Espigão Mestre, 6-16-III-1972, *W. R. Anderson et al. 37083, 36701* (NY).

Ceará. *Loefgren 644* (R); Chapada do Araripe, *Allemão & Cysneiros 49* (R), 3-I-1949, *Andrade-Lima 49-160* (IPA).

Goiás. *Burchell 6987* (K), *8679* (K); Brasilia, D. F., 20-II-1964, *Labouriau & Valio 1190* (SP), 14-VIII-1964, *Irwin & Soderstrom 5125* (UB), 9-IX-1965, *Irwin et al. 11128* (UB), 13-XII-1965, *11255* (UB), 29-I-1966, *12101* (UB), 4-V-1966, *15608* (UB), 3-XI-1969, *Langenheim 5651, 5655, 5657* (UC), 9-I-1972, *Lee 141, 142* (UC); Cristalina and its vicinity, 5-XI-1965, *Irwin et al. 9969* (UB), 3-III-1966, *13333* (UB), 7-III-1966, *13673* (UB), 8-III-1966, *13762* (UB); 50 km. to Caiaponia, on road to Jataí, 27-VI-1966, *17853* (UB); Formosa, 22-IV-1966, *15269* (UB), 29-IV-1966, *15484* (NY, UC); Rio Corumbá, near Luziânia, 5-XI-1969, *Langenheim 5662, 5663, 5664, 5665* (UC); Serra dos Pireneus, 15-I-1972, *Irwin et al. 34053, 34150, 34173, 34174, 34175, 34176* (NY, UB, UC), *34607* (NY); Niquelândia, 21-I-1972, *34732* (NY, UB, UC), *Lee 129* (UC).

Mato Grosso. 16-XII-1967, *D. Philcox et al. 3453* (UB); Rio Juruena, *B. Maguire et al. 56483* (NY, UB); 20 km. past Alto Araguaia en route to Alta Garça, *56957* (F, NY); Campo Grande, 20-II-1952, *Kuhlmann SP-69965* (SP), *RB-57182* (RB), 6-IX-1936, *Archer & Gehut 121* (SP, US); Xavantina and its vicinity, 30-V-1966, *Irwin et al. 16308* (UB), 4-VI-1966, *16569* (UB), 12-VI-1966, *17510* (UB).

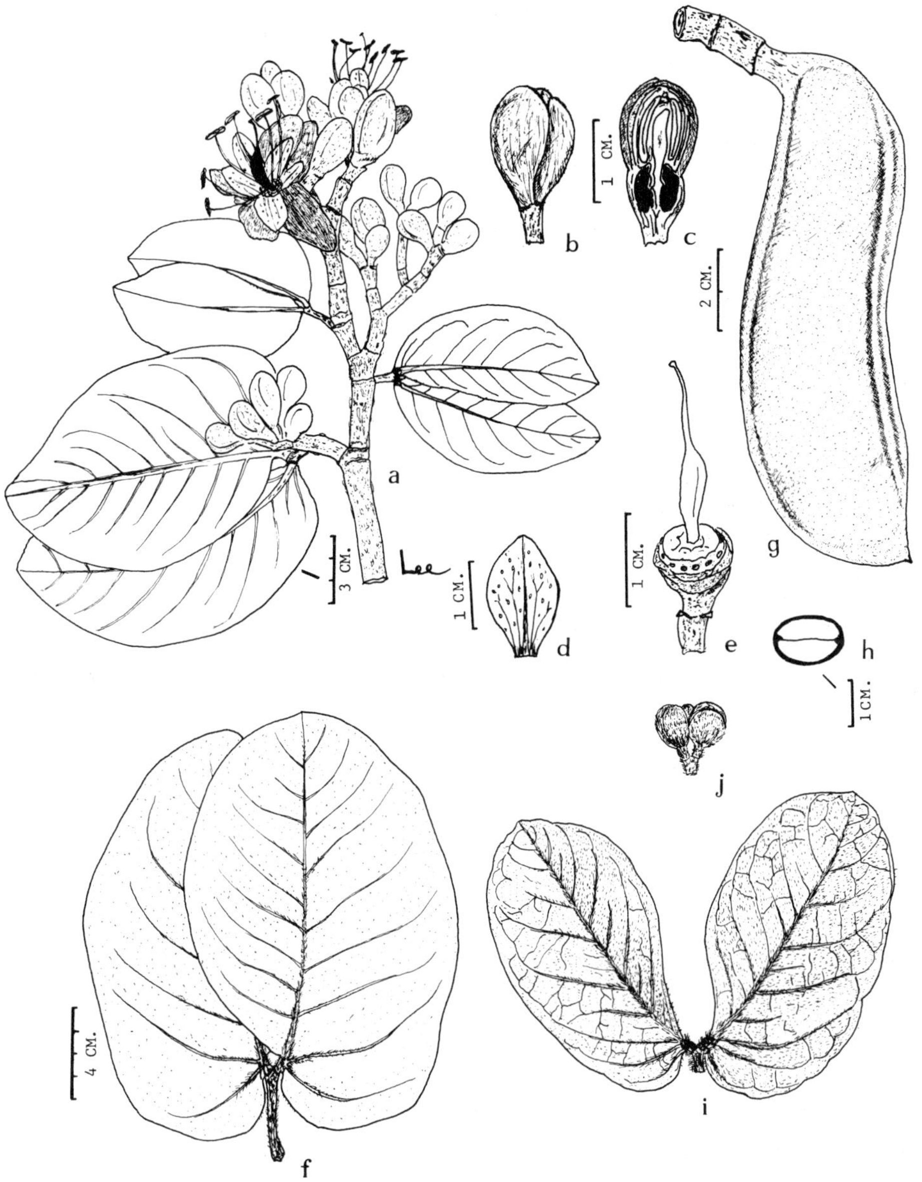

32. *Hymenaea stigonocarpa.* a-e, var. *stigonocarpa:* a. flowering branch. b. flowering bud (note the two bracteoles). c. longitudinal section of a flowering bud showing mass of nectariferous tissue (darkened portion). d. petal. e. pistil, hypanthium, and pedicel (note the elevated disc). f-h, var. *pubescens:* f. mature leaf (abaxial side). g. fruit. h. seed (cross-section). i-j, var. *brevipetiolata:* i. mature leaf (abaxial side). j. flowering buds. i and j redrawn from Mattos, 1964.

Minas Gerais. *M. Claussen 793* (A, F); *Blanchet 3154* (P); Mun. Várzea da Palma, 26-II-1964, *Labouriau & Valio 1241* (SP); Paraopeba, 22-III-1956, *Herringer* (MG, UB); Prata, 20-III-1963, *G. M. Feleppe 152* (RB, SP, US); Lagoa Santa, Santa Luzia, *M. Barreto 4407* (R, SP); 203 km. S of Paracatu, on road to Patos, 9-II-1959, *Irwin 2616* (F, NY, R, UC, US); Coromandel, 25-VII-1947, *R-111964* (R); Tres Barras, *L. B. Smith 6909* (P).

Pará. Serra do Cachimbo, XII-1956, *Pires* (IPEAN).

São Paulo. Moji-Guaçu, X-1962, *J. Mattos 10566* (SP), V-1954, *Kuhlmann 4199* (SP); 26-XI-1969, *Langenheim 5671* (UC); Itirapina, São Paulo. São Carlos, *N. D. Cruz 124* (SP).

PARAGUAY. "In regione cursus sperioris fluminis Apa," *Hassler 7882* (A, NY, P); between Rio Apa and Rio Aquidaban, 31-XII-1908/1909, *K. Fieberg 4523* (A).

BOLIVIA?

Vernacular names: "Jatobá de chapada" (Piauí), "jatobá do cerrado" (Goiás), and "jatobá do campo" (Mato Grosso).

SPECIES DUBIA

Hymenaea travassii Kuhl. ex Paes, Rodriguesia 26(38): 1. 1971

Type: Bolivia, Aguas Calientes, *L. Travassos EB 135561* (RB, holotype).

Arbor parva ramis foliisque glabris, nitidis, foliolis basi assymetricis, apice obtusis, sessilibus, translucidis, laminis ellipticis, utrinque pallide viridibus, subglaucis, facie antica evidentibus et parvi-reticulatis, postica obscure reticulatis, densiuscule nigro-punctatis et glandulis sparsis ad basin notatis, 3.5-6.8 cm. longis, 2.5-3.5 cm. latis; raque pilosulis-pedunculis brevissimis pilosis; Alabastra dense griseo pilosa, pedunculis parvis glabris, pedicellis nullis vel brevissimis, rotundata, bractea extus pilosa, intus glabra, calice segmentis extus valde pilosis et intus sericeo-pilosis, valde imbricatis, concavis, submembranaceis, petalis glabris, evidente translucido-pilosis, ovario glabro. Fructus unicus, non bene evolutus, dense ruguloso-glandulosus, basi breve stipitatus, apice rudimentis styli coronatus, oblongo-ellipticus plus minus assymetricus, 4 cm. longus et 1.7 cm. latus. Paes (1971, p. 1).

This species is known only from the type collection. We have been informed by the Director of Jardim Botânico do Rio de Janeiro, Brazil, that the type could not be located. From the description and the accompanying illustration of a branch with young flowering buds and a single immature fruit, *H. travassii* appears sufficeintly closely related to *H. courbaril* to not deserve species status.

NOMINA NUDA

1. *Hymenaea copaeifera* Mart. ex Hayne, nomen, Arzneik. gebräuchl. Gewächses 11: pl. 15. 1830.
2. *Hymenaea hayneaea* D. Dietr., nomen, Syn. Pl. 2: 1473. 1840.
3. *Hymenaea microcarpa* Huber, nomen, Bol. Mus. Paraense Hist. Nat. 6: 209, 1910.
4. *Hymenaea microphylla* B. Rodr., nomen, Hort. Flumin. 149. 1895.
5. *Hymenaea mirabilis* Saldanha, nomen, da Gama, Config. & Descr. Madier, Rio de Janeiro. 118. 1865.
6. *Hymenaea pororoca* Huber, nomen, Bol. Mus. Paraense Hist. Nat. 6: 209. 1910.
7. *Hymenaea pubescens* Mart. ex Hayne, nomen, Arzneik. gebräuchl. Gewächses 11: pl. 16. 1830.

LITERATURE CITED

Adams, R., and B. L. Turner
1970. Chemosystematic and numerical studies of natural populations of *Juniperus ashei.* Buch. Taxon 19 (5):728-751.

Allen, G. M.
1939. Bats. Cambridge, Massachusetts, Harvard University Press.

Amshoff, G. J. H.
1939. Caesalpinioideae. *In* A. Pulle (ed.), Flora of Suriname (Netherlands Guiana) 2(2): 23.

Andrade-Lima, D.
1953. Notes sôbre a dispersão de algunas especies vegetais no Brasil. Ann. Soc. Biol. Pernambuco 11(1): 25-49.

Andrew, E. C.
1915. The development and distribution of the natural order Leguminosae. J. & Proc. Roy. Soc. New South Wales 1914: 333-407.

Araujo, V. C. de
1970. Fenologia de essências florestais Amazônicas 1. Bol. INPA. Pesquisas Florestais 4: 1-25.

Ashton, P. S.
1969. Speciation among tropical forest trees: some deductions in the light of recent evidence. Biol. J. Linn. Soc. 1: 155-196.

Atchison, E.
1951. Studies in Leguminosae, VI: Chromosome numbers among tropical woody species. Amer. J. Bot. 38: 538-546.

Baillon, H.
1870. Histoire des Plantes, 2: 114. Paris.

Baker, E. G.
1926. The Leguminosae of Tropical Africa. Ghert, Erasmus Press.

Baker, G. J.
1877. Flora of Mauritius and the Seychelles. London, L. Reeve & Co.

Baker, H. G.
1959. Reproductive methods as factors in speciation in flowering plants. Cold Spring Harbor Symp. Quant. Biol. 24: 177-191.
1961. The adaptation of flowering plants to nocturnal and crepuscular pollinators. Bot. Rev. 36: 277-319.
1963. Evolutionary mechanisms in pollination biology. Science 139: 877-883.
1970. Two cases of bat pollination in Central America. Rev. Biol. Trop. 17: 187-197.
1973. Evolutionary relationships between flowering plants and animals in American and African tropical forest. *In* B. J. Meggers, E. S. Ayensu, and W. D. Duckworth (eds.), Tropical Forest Ecosystems in Africa and South America: A Comparative Review, 145-159, Washington, D.C., Smithsonian Press.

Balser, C.
1960. Notes on resin in aboriginal Central America. Akt. 34. Int. Amerik. Kong. Wien. 374-380.

Bauhin, C.
1623. Pinax Theatri Botanici, 403.

Bawa, K. S.
1973. Chromosome numbers of tree species of a lowland tropical community. J. Arnold Arbor. 54(4): 422-434.
1974. Breeding systems of tree species of a lowland tropical community. Evolution 28(1): 85-92.

Bawa, K. S., and P. A. Opler
1975. Dioecism in tropical forest trees. Evolution 29. In press.

Bentham, G.
1840. IV. Contributions toward a flora of South America – Enumeration of plants collected by Mr. Schomburgk in British Guiana. Hooker's J. Bot. Kew Gard. Misc. 2: 38-146.
1870. Leguminosae II. Swartzieae et Caesalpinieae. *In* Martius' Flora Brasiliensis 15(2): 234-237.

Bentham, G., and J. D. Hooker

1865. Genera Plantarum, 1(2): 562-600.
Berger, C. A., E. R. Witkus, and R. M. McMahon
1958. Cytotaxonomic studies in the Leguminosae. Bull. Torrey Bot. Club 85: 405-415.
Black, G., Th. Dobzhansky, and C. Pavan
1950. Some attempts to estimate species diversity and population density of trees in Amazonian forests. Bot. Gaz. (Crawfordsville) 3: 413-425.
Brenan, J. P. M.
1965. The geographical relationships of the genera of the Leguminosae in tropical Africa. Webbia 19(2): 545-578.
1967. Leguminosae – subfamily Caesalpinioideae. *In* E. Milne-Redhead and R. Polhill, Flora of Tropical East Africa. Crown Agents for Overseas Government and Administration.
Breyne, J.
1679. Prodromus Fasciculi Rariorum Plantarum, 2: 8.
Britton, N. L., and J. N. Rose
1930. Caesalpiniaceae. *In* North American Flora 23(4): 224.
Burkart, A.
1952. Les Leguminosas Argentinas, 2nd. ed. Buenos Aires, Acme Agency S.R.L.
Candolle, A. P. de
1825a. Prodromus Systematis Naturalis Regni Vegetabilis, 2: 94. Paris.
1825b. Memoires sur la Famille des Legumineuses. Paris.
Carvalho, C. T. de
1960. Das visitas de Morcegos às Flôres (Mammalis, Chiroptera). Anais Acad. Brasil. Ci. 32: 359-377.
1961. Sobre as hábitos alimentares de Phillostomideos (Mammalia, Chiroptera). Rev. Bio. Trop. 9(1): 53-60.
Celakovský, L. F.
1892. Gedanken uber eine zeitgemässe Reform der Theorie der Blütenstande. Bot. Jahrb. Syst. 16: 33-51.
Cordemoy, H. J.
1900. Les Gommes et les Resines. Ann. Inst. Bot. – Geól. Colon. Marseille 7: 1-45.
Corner, E. J. H.
1954. The evolution of tropical forest. *In* J. S. Huxley, J. S. Hardy, A. C. and E. B. Ford (eds.), Evolution as a Process, London, Allen & Unwin.
Cowan, R. S.
1953. A taxonomic revision of the genus *Macrolobium* (Leguminosae – Caesalpinioideae). Mem. New York Bot. Gard. 8(4): 257-342.
Cuatracasas, J.
1957. A sketch of the vegetation of the North Andean Province. Proc. 8th Pacific Science Congr. 4: 167-173.
Cunningham, A., S. S. Martin, and J. H. Langenheim
1973. Resin acids from two Amazonian species of *Hymenaea*. Phytochemistry 12: 633-635.
1974. Labd-13-en-8-o1-15-oic acid in trunk resin of Amazonian *Hymenaea courbaril*. Phytochemistry 13: 294-295.
Dietrich, D.
1840. Synopsis Plantarium, 2: 1473.
Dobzhansky, Th.
1950. Evolution in the tropics. Amer. Sci. 38: 209-221.
Ducke, A.
1925. As Leguminosas do Estado do Pará. Arch. Jard. Bot. Rio de Janeiro 4: 263.
1935. As especies Brasileiras de jatahy, jutahy ou jatobá (genero *Hymenaea* L., Leguminosas Caesalpiniaceas). Ann. Acad. Brasil Sci. 7: 203-211.
1938. Notes on the purpleheart woods of Brazilian Amazonia. Trop. Woods 54: 1-7.
1949. As Leguminosas da Amazonia Brasileira, 2nd ed. Bol. Técn. Inst. Agron. N. 18: 1-248.
Ducke, A., and G. A. Black
1953. Phytogeographical notes on the Brazilian Amazon. Ann. Acad. Brasil. Sci. 24(1): 1-46.

Dwyer, J.
1954. Rapport entre stipe et coupe réceptaculaire dans la classification des Amherstieae. Proc. VIII Int. Bot. Congr., Paris.
1958. The New World species of *Cynometra.* Ann. Missouri Bot. Gard. 45: 303-345.

Egler, W.
1963. Adolpho Ducke – Tracos biográficos, viagens e trabalhos. Bol. Mus. Paraense Hist. Nat. 18: 51.

Eiten, G.
1972. The cerrado vegetation of Brazil. Bot. Rev. 38(2): 201-341.

Engler, A.
1915. *In* A. Engler and O. Drude, Die Pflanzenwelt Afrikas. 3(1): 450-451. Leipzig.

Fasbender, M. V.
1959. Pollen grain morphology and its taxonomic significance in the Amherstieae, Cynometreae, and Sclerolobieae (Caesalpiniaceae) with special reference to the American genera. Lloydia 22(2): 107-162.

Federov, An. A.
1966. The structure of the tropical rain forest and speciation in the humid tropics. J. Ecol. 54:1-11.

Flake, R. H., and B. L. Turner
1968. Numerical classification for taxonomic problems. J. Theoret. Biol. 20: 260-270.

Gaertner, J.
1791. De Fructibus et Seminibus Plantarum, 2: 305-306. Stutgardial.

Gilg, E.
1897. VI.Über die Stammpflanzl des Zanzibar-Kopals. Notizbl. Königl. Bot. Gart. Berlin. 1: 198-205.

Goebel, K. I. E.
1931. Blutenbildung und Sprossgestaltung. Anthokladian und Infloreszenzen. Suppl. 2 to Organographie der Pflanzen, 3rd ed.

Grant, V.
1956. The influence of breeding habit on the outcome of natural hybridization in plants. Amer. Naturalist 90: 379-389.

Guignard, L.
1892. L'appareil sécréteur de *Copaifera.* Bull. Soc. Bot. France 39: 253.

Gunn, R.
1968. Stranded seeds and fruits from southeastern shore of Florida. Garden J. (March/April) 43: 54.

Haffer, J.
1969. Speciation in Amazonian forest birds. Science 163: 1131-1136.

Harborne, J. B.
1971. Terpenoid and other low-molecular weight substances of systematic interest in the Leguminosae. *In* J. B. Harborne, D. Boulter, and B. L. Turner (eds.), Chemotaxonomy of the Leguminosae, 257-283, New York, Academic Press.

Harms, H.
1915. Caesalpinioideae. *In* A. Engler & O. Drude, Vegetation de Erde 93(1): 423-520.

Hasskarl, J. K.
1842. Plantarum genera et species novae aut reformatae javenses. Flora 25: 95-97.

Hayne, F. G.
1827. Ueber die Gattungen *Hymenaea, Vouapa* und eine neue *(Trachylobium)* mit hinsicht auf die Abstammung des aus Amerika kommenden Copals. Flora 10: 737-746.
1830. Arzneikunde gebräuchlichen Gewächses XI: 6-16. Berlin.

Heinrich, B., and P. H. Raven
1972. Energetics and pollination ecology. Science 176: 597-602.

Heithaus, E. R., T. H. Fleming, and P. A. Opler
1974. Foraging pattern and resource utilization by eight species of bats in a seasonal tropical forest. Unpublished manuscript.

Heywood, V. H.
1971. The Leguminosae – a systematic purview. *In* J. B. Harborne, D. Boulter, and B. L. Turner (eds.), Chemotaxonomy of the Leguminosae, 1-29, New York, Academic Press.

Hellinckx, L.
1955. Les propriétés des copals du Congo Belge en relation avec leur origine botanique. Institut National pour l'Etude Agronomique de Congo Belge. Publications Service no. 44, Bruxelles: 8-25.
Hickey, L. J.
1973. Classification of architecture of dicotyledonous leaves. Amer. J. Bot. 60: 17-33.
Holdridge, L. R.
1964. Life Zone Ecology. Tropical Science Center, San José, Costa Rica.
Holdridge, L. R., W. C. Grenke, W. H. Hatheway, T. Liang, and J. A. Tosi, Jr.
1971. Forest Environments in Tropical Life Zones: a Pilot Study. New York, Pergamon Press.
Howes, F. N.
1949. Vegetable Gums and Resins. Chron. Bot. Waltham, Mass.
Huber, J.
1909. VI. Plantae Duckeanae austro-guyanenses. Mattas e madeiras amazonicas. Bol. Mus. Paraense Hist. Nat. 5: 385-386.
Hugel, G., A. C. Oehlschlager, and G. Ourisson
1966. The structure and stereochemistry of diterpenes from *Trachylobium verrucosum* Oliv. Tetrahedron Suppl. 8, Pt. I: 203-216.
Hugel, G., L. Lods, J. M. Mellor, D. W. Theobald, and G. Ourisson
1965a. Diterpènes de *Trachylobium*. I. Introduction générale. Isolement du kauranol et de huit diterpènes. Bull. Soc. Chim. Fr. 1965: 2882-2887.
1965b. Diterpènes de *Trachylobium*. II. Structure des diterpènes têtra- et pentacycliques de *Trachylobium*. Bull. Soc. Chim. Fr. 1965: 2888-2894.
Hugel, G., L. Lods, J. M. Mellow, and G. Ourisson
1965. Diterpènes de *Trachylobium*. III. Reactions des dérivés trachylobaniques. Bull. Soc. Chim. Fr. 1965: 2894-2902.
Hugel, G., and G. Ourisson
1965. Diterpènes de *Trachylobium*. IV. Structure et stereochimie de l'acide zanzibarique. Bull. Soc. Chim. Fr. 1965: 2903-2908.
Hutchinson, J.
1964. The Genera of Flowering Plants. I. Dicotyledons. Oxford, Clarendon Press.
Janzen, D. H.
1967. Synchronization of sexual reproduction of trees within the dry season in Central America. Evolution 21: 620-637.
1969. Seed-eaters versus seed size, number, and dispersal. Evolution 23: 1-27.
1970. Herbivores and the number of tree species in tropical forests. Amer. Naturalist 104: 501-528.
1971a. Euglossine bees as long distance pollinators of tropical plants. Science 171: 203-205.
1971b. *Scheelea, Hymenaea,* and *Sickingia* and their seed predators. Organization for Tropical Studies, News 71(5): 3-4.
Jussieu, A. L. de
1789. Genera Plantarum.
Khoo, S. F., A. C. Oehlschlager, and G. Ourisson
1973. Structure and stereochemistry of diterpenes of *H. courbaril* (Caesalpinioideae) seed pod resin. Tetrahedron 29: 3379-3388.
Kirk, J.
1868. On the copal of Zanzibar. J. Linn. Soc. Bot. 11: 1-4.
Kleinhoonte, A.
1926. Leguminosae. *In* A. Pulle, Neue Beitrage zur Flora Surinams IV, Recueil Trav. Bot. Neerl. 22: 391-417.
Klotzsch, F. G.
1862. *In* W. C. H. Peters, Reise nach Mossambique Botanik, 1: 22, pl. 2.
Knaap-Van Meeuwen, M. S.
1970. A revision of four genera of the tribe Leguminosae-Caesalpinioideae-Cynometreae in Indomalesia and the Pacific. Blumea 18(1): 1-52.

Kooiman, P.
1960. On the occurrence of amyloids in plant seeds. Acta Bot. Neerl. 9: 208-219.
Kunth, C. S.
1825. Nova Genera et Species Plantarum: 254-255, pls. 566-567.
Lamarck, J. B. A. P. M.
1792. Tableau Encyclopedique et Methodique des Trois Regnes de la Nature (Illustration des genres) 2: 825, pl. 330.
Langenheim, J. H.
1966. Botanical source of amber from Chiapas, Mexico. Ciencia (Mexico) 24: 201-211.
1967. Preliminary investigations of *Hymenaea courbaril* as a resin producer. J. Arnold Arbor. 48: 203-230.
1968. Infrared photometric study of resins from northeast Angola. Appendix. Observations on Forest Destruction in the Congo Basin in Prehistoric Times with Special Reference to Northeast Angola. Publicacões Culturais, Companhia de Diamantes de Angola 78: 151-154.
1969. Amber: A botanical inquiry. Science 163: 1157-1169.
1972. Origin of fossilized resin and its relation to forest history in northeastern Angola. Publicacões Culturais, Companhia de Diamantes de Angola 85: 15-36.
1973. Leguminous resin-producing trees in Africa and South America. *In* B. J. Meggers, E. S. Ayensu, and W. D. Duckworth (eds.), Tropical Forest Ecosystems in Africa and South America: A Comparative Review, 89-104, Washington, D.C., Smithsonian Press.
Langenheim, J. H., and C. W. Beck
1965. Infrared spectra as a means of determining botanical source of amber. Science 149: 52-55.
1968. Catalogue of infrared spectra of amber. Pt. I. North and South America. Harvard Bot. Mus. Leaflets 22: 65-120.
Langenheim, J. H., and Y. T. Lee
1974. Reinstatement of the genus *Hymenaea* L. (Leguminosae: Caesalpinioideae) in Africa. Brittonia 26: 3-21.
Langenheim, J. H., Y. T. Lee, and S. S. Martin
1970. Amazonian species of the resin producing genus *Hymenaea.* Amer. J. Bot. 57: 754.
1971. Brazilian species of the leguminous resin-producing genus *Hymenaea.* Amer. J. Bot. 58: 466.
1973. An evolutionary and ecological perspective of the Amazonian hylaea species of *Hymenaea* (Leguminosae: Caesalpinioideae). Acta Amazonica III(1): 5-37.
Lanjouw, J., and F. A. Stafleu
1964. Index Herbariorum. 5th ed. Int. Bur. Pl. Tax. and Nomencl., Utrecht, Netherlands, Regnum Vegetabile: 31.
Le Cointe, P.
1947. Amazonia Brasileira, III: Arbores e plantas úteis (Indígenas e aclimadas), 2nd ed. São Paulo, E. Nacional p. 46, 47, 193, 250, 379, Brasiliana, série 5.a, 251.
Lee, Y. T., and J. H. Langenheim
1973. New taxa from Brazil and Guyana in the genus *Hymenaea* (Leguminosae, Caesalpinioideae). J. Arnold Arbor. 54(1): 94-104.
1974a. Additional new taxa and new combinations in *Hymenaea* L. (Leguminosae, Caesalpinioideae). J. Arnold Arbor. 55(3): 441-452.
1974b. *In* Á. Löve's IOPB Chromosome number report XLV (Chromosome number of *Hymenaea* L.). Taxon 23(4): 621-622.
León, El Hermano
1949. El genero *Hymenaea* en Cuba. Revista Soc. Cub. Bot. 6: 1-7.
Léonard, J.
1951. Les *Cynometra* et les genres voisins en Afrique tropical: Notulae Systematicae XI. Bull. Jard. Bot. État. 21: 373-450.
1957. Genera des Cynometreae et des Amherstieae Africaines. Mem. Cl. Sci. Acad. Roy. Belgique 30(2): 1-314.
Lima, P. F.
1950. A canoa de casca de jatobá entre os indios do Xingu. Revista, São Paulo, Brasil Museu Paulista. N.S. 4: 369-380.

Linnaeus, C. von
1737a. Genera Plantarum. Leiden.
1737b. Hortus Cliffortianus. Amsterdam.
1753. Species Plantarum. Stockholm.
Marcgraf, G.
1648. Historia Naturalis Brasilliae (São Paulo Imprensa Official do Estado. 1942).
Martin, S. S.
1973. Characterization and comparison of sesquiterpenes in leaf pocket resins of *Hymenaea* L. (Leguminosae, Caesalpinioideae). PhD dissertation, University of California, Santa Cruz.
Martin, S. S., and J. H. Langenheim
1974. Enantio-8(17),13(16),14-labdatrien-18-oic acid from trunk resin of Kenyan *Hymenaea verrucosa.* Phytochemistry 13: 523-525.
Martin, S. S., J. H. Langenheim, and E. Zavarin
1972. Sesquiterpenes in leaf pocket resin of *Hymenaea courbaril.* Phytochemistry 11: 3049-3051.
1973. Compositional variation of resin pocket sesquiterpenes in *Trachylobium verrucosa.* Biochem. Syst. 1: 35-37.
1974. Quantitative variation in leaf pocket resin composition in *Hymenaea courbaril* (Leguminosae). Biochem. Syst. and Ecol. 3: 75-87.
Martinez, M.
1960. Una especie de *Peltogyne* en México. Anales Inst. Biol. Univ. Nac. México 31: 123-131.
Mattos, N. F.
1968. Estudos sistemáticos no gênero *Hymenaea.* Dusenia 8(5): 153-160.
McArthur, R. H.
1969. Patterns of communities in the tropics. Biol. J. Linn. Soc. London, I: 19-30.
Melchior, H.
1964. A. Engler's Syllabus der Pflanzenfamilien, 12th ed. Vol. 2. Gebrüder Borntraeger, Berlin Nikolassee.
Menendéz, C. A.
1969. Die Fossilen Floras Sudamerikas. *In* E. F. Fittkau *et al.*, Biogeography and Ecology in South America 2: 519-561, Junk, The Hague.
Miquel, F. A. W.
1855. Flora Indiae Batavae 1(1): 81-82. Leipzig.
Miranda, F., and E. Hernandez-X
1963. Los Tipos de vegetación de México y su clasificación. Bol. Soc. Bot. Méx. 28: 29-179.
Mirov, N.
1967. The Genus *Pinus.* New York, Ronald Press.
Moens, P.
1955. Les formations secretrices des copaliers. Cellule 57: 35-59.
Moreau, R. E.
1966. The Bird Faunas of Africa and its Islands. New York, Academic Press.
Mors, W. B.
1946. A hemicelulose das sementes de *Hymenaea parvifolia* Huber e seu emprego na cremagem do latex de seringueira. Bol. Tecn. Inst. Agron. N. 6: 1-42.
Nägeli, C. W. von
1884. Mechanish – Physiologische Theorie der Abstammungslehre. Munchen.
Nakano, T., and C. Djerassi
1961. Terpenoids XLVII. Copalic acid. J. Org. Chem. 26: 167-173.
Nazif, M. J.
1914. Sur le fruit des *Hymenaea.* Bull. Soc. Bot. Geneve 11: 29-32.
Noriega, J. M.
1918. El Guapinole. Bol. Direccion Etud. Biol. 2: 357-363.
Oliver, D.
1871. Flora of Tropical Africa, 2. London, L. Reeve & Co.
Paes, L. E.
1971. *Hymenaea travassii* Kuhlmann. Rodriguesia 26(38): 1-2.

Pilger, R. K. F.
1922. Uber Verzweigung und Blutenstandsbildung bei den Holzgewachsen. Bibl. Bot. 23: 1-38.
Pires, J. M., Th. Dobzhansky, and G. Black
1953. An estimate of the number of species in an Amazonian forest community. Bot. Gaz. (Crawfordsville) 6: 467-477.
Piso, G.
1642. Historia, Natural e Médica da Índia Ocidental. Reprint, Rio de Janeiro, 1957.
Plukenet, L.
1699. Almagestum Botanicum, 2: 96, t. 82, fig. 2.
Plumier, C.
1703. Nova Plantarum Americanarum Genera. Paris.
Ponsinet, G., G. Ourisson, and A. C. Oehlschlager
1968. Systematic aspects of the distribution of diterpenes and triterpenes. *In* T. J. Mabry, R. E. Alston, and V. C. Runeckles (eds.), Recent Advances in Phytochemistry, 1: 271-302, New York, Appleton-Century-Crofts.
Poore, M. E. D.
1968. Studies in Malaysian rain forest I. The forest on Triassic sediments in Jengka Forest Reserve. J. Ecol. 56: 143-196.
Prance, G. T.
1974. Phytogeographic support for the theory of Pleistocene forest refuges in the Amazon Basin, based on evidence from distribution patterns in Caryocaraceae, Chrysobalanaceae, Dichapetulaceae, and Lecythidaceae. Acta Amazonica III(3): 5-26.
Ratter, J. A., P. W. Richards, G. Argent, and D. R. Gifford
1973. Observations on the vegetation of northeastern Mato Grosso. 1. The woody vegetation types of the Xavantina-Cachimbo expedition area. Philos. Trans. Ser. B. 266 (no. 880): 449-492.
Raven, P. R., and D. I. Axelrod
1972. Plate tectonics and Australasian paleobiogeography. Science 176: 1379-1386.
Ray, J.
1688. Historia Plantarum, 2: 1760.
Record, S. J., and R. W. Hess
1943. Timbers of the New World. New Haven, Yale University Press.
Record, S. J., and C. D. Mell
1924. Timbers of Tropical America. New Haven, Yale University Press.
Rickett, H. W.
1944. The classification of inflorescences. Bot. Rev. 10: 187-231.
Rizzini, C. T.
1963. Nota prévia sobre a devisão fitogeográfica do Brasil. Revista Brasil. Geogr. 25(1): 3-64.
1967. Delimitacão e relacoẽs da flora silvestre hileiana. *In* H. Leng (ed.), Atlas do Simposio sôbre a Biota Amazonica, 4(Botânica): 13-36. Conselho Nacional de Pesquisas, Rio de Janeiro.
Sandermann, W.
1962. Terpenoids: metabolism. *In* M. Florken and H. S. Mason (eds.), Comparative Biochemistry, 591-630, New York, Academic Press.
Sandwith, N. Y.
1931. XLVIII. Contributions to the flora of tropical America: VII. New and noteworthy Leguminosae and Rosaceae from British Guiana. Bull. Misc. Inf. Roy. Bot. Gard. Kew 1931: 366-367.
1938. New species and records from British Guiana: XV. Contributions to the flora of tropical America: XXX. Bull. Misc. Inf. Roy. Bot. Gard. Kew 1937(2): 100-112.
Schultes, R. E.
1953. Plantae Austro-Americanae VIII de plantis principaliter callis Amazonicis novis vel criticis notae diversae. Harvard Bot. Mus. Leaflet 16(4): 57-85.
Simpson, G. G.
1969. South American mammals. *In* E. F. Fittkau *et al.*, Biogeography and Ecology in South America 2: 879-909, Junk, The Hague.

Smith-White, S.
1959. Cytological evolution in the Australian flora. Cold Spring Harbor Symp. Quant. Biol. 24: 273-289.
Snow, D. W.
1966. A possible selective factor in the evolution of flowering seasons in tropical forest. Oikos 15: 274-281.
Sokal, R. R., and P. H. A. Sneath
1963. Principles of Numerical Taxonomy. San Francisco, W. H. Freeman Co.
Sornay, P. de
1913. Les Plantes Tropicales de la Famille de Legumineuses.
Stebbins, G. L.
1957. Self fertilization and population variability in higher plants. Amer. Naturalist 91: 337-354.
1958. Longevity, habitat, and release of genetic variability in higher plants. Cold Spring Harbor Symp. Quant. Biol. 23: 365-378.
1970. Adaptive radiation of reproductive characteristics in Angiosperms. I. Pollination mechanisms. Ann. Rev. Ecol. Syst. 1: 307-326.
1971. Adaptive radiation of reproductive characteristics in Angiosperms. II. Seeds and seedlings. Ann. Rev. Ecol. Syst. 2: 237-260.
Stein, O.
1975. The shoot apex of *Hymenaea courbaril* L. Amer. J. Bot. 62: 303-310.
Taubert, P.
1891. Leguminosae. *In* A. Engler and K. Prantl, Die naturlichen Pflanzenfamilien 3(3): 70.
Thorne, R. F.
1973. Floristic relationships between tropical Africa and tropical America. *In* B. J. Meggers, E. S. Ayensu, and W. D. Duckworth (eds.), Tropical Forest Ecosystems in Africa and South America: A Comparative Review, 27-47, Washington, D. C., Smithsonian Press.
Tschirch, A., and E. Stock
1936. Die Hazar. 3rd ed. Berlin.
Turner, B. L., and D. S. Fearing
1959. Chromosome numbers in the Leguminosae. II: African species, including phyletic interpretations. Amer. J. Bot. 46: 49-57.
Vahl, M.
1792. Ecologie Americanae, 2: 31.
Van der Hammen, T.
1961. The Quaternary climatic changes of northern South America. New York Acad. Sci. Annal 95: 676-683.
1972. Changes in vegetation and climate in the Amazon Basin and surrounding areas during the Pleistocene. Geologie en Mijnboua 50(6): 641-643.
Van der Pijl, L.
1936. Fledermäuse und Blumen. Flora 131: 1-40.
1956a. Remarks on pollination by bats in the genera *Freycinetia, Duabanga* and *Haplophragma,* and on chiropterophily in general. Acta Bot. Neerl. 5: 135-144.
1956b. Classification of the leguminous fruits according to their ecological and morphological properties. Proc. Akad. Westenschappen, ser. C, 56: 301-313.
Vanzolini, P. E.
1973. Paleoclimates, relief and species multiplication in equatorial forests. *In* B. J. Meggers, E. S. Ayensu, and W. D. Duckworth (eds.), Tropical Forest Ecosystems in Africa and South America: A Comparative Review, 225-258, Washington, D.C., Smithsonian Press.
Vanzolini, P. E., and E. E. Williams
1970. South American anoles: the geographic differentiation and evolution of the *Anolis chrysolepis* species group (Sauria, Iguanidae). Arq. Zool. São Paulo 19: 1-240.
Veevers, J. H., J. G. Jones, and J. A. Talent
1971. Indo-Australian stratigraphy and the configuration and dispersal of Gonwanaland. Nature 229: 383-388.

Vogel, S.
1968-1969. Chiropterophilie in der neotropischen Flora. Neue Mitteilungen. I, II, and III. Flora Abt. B 157: 562-602; 158: 185-222, 289-323.
Vogel, T.
1837. De Caesalpinieis Brasiliae. Linnaea 11: 409-410.
Vuilleumier, B. S.
1971. Pleistocene changes in the flora and fauna of South America. Science 173: 771-780.
Wolfe, J. A.
1971. Tertiary climatic fluctuations and methods of analysis of Tertiary flora. Palaeogeogr. Palaeoclimatol. Palaeoecol. 9: 27-57.
Woodson, R. E., and R. W. Schery
1951. Leguminosae, 2(3). Flora of Panama. Ann. Missouri Bot. Gard. 38: 301-394.
Zavarin, E.
1971. Monoterpenoids of coniferales, 57-63. Biochemical Interactions among Plants, National Academy of Sciences, Washington, D.C.
Zavarin, E., and E. Frank
1965. Chemical composition of *P. elliottii* var. *elliottii* turpentine and its possible relation to taxonomy of several pine species. Phytochemistry 4(4): 563-568.
Zavarin, E., and K. Snajberk
1973. Geographic variability of monoterpenes from cortex of *Pseudotsuga menziesii.* Pure and Applied Chem. 34: 411-434.
Zavarin, E., N. T. Mirov, and K. Snajberk
1966. Chemical composition of the turpentine of some eastern Mediterranean pines in relation to their classification. Phytochemistry 5(1): 97-102.